A History of Data Visualization and Graphic Communication

A History of Data Visualization and Graphic Communication

Michael Friendly
Howard Wainer

HARVARD UNIVERSITY PRESS
Cambridge, Massachusetts, and London, England · 2021

Printed in the United States of America

Second printing

Library of Congress Cataloging-in-Publication Data

Names: Friendly, Michael, author. | Wainer, Howard, author.
Title: A history of data visualization and graphic communication / Michael Friendly, Howard Wainer.
Description: Cambridge, Massachusetts : Harvard University Press, 2021. | Includes bibliographical references and index.
Identifiers: LCCN 2020047837 | ISBN 9780674975231 (hardcover)
Subjects: LCSH: Information visualization—History. | Visual communication—History. | Graphic methods—History. | Visual analytics—History.
Classification: LCC QA76.9.I52 F74 2021 | DDC 001.4/226—dc23
LC record available at https://lccn.loc.gov/2020047837

To Martha,
Abigail and Gustavo, Ethan,
Luca and Oliver

To Linda,
Sam,
Laurent, Lyn, Koa and Sophie

Contents

A History of Data Visualization
and Graphic Communication

Introduction

The only new thing in the world is the history you don't know.

—HARRY S. TRUMAN, quoted by David McCulloch

We live on islands surrounded by seas of data. Some call it "big data." In these seas live various species of observable phenomena. Ideas, hypotheses, explanations, and graphics also roam in the seas of data and can clarify the waters or allow unsupported species to die. These creatures thrive on visual explanation and scientific proof. Over time new varieties of graphical species arise, prompted by new problems and inner visions of the fishers in the seas of data.

Whether we're aware of this or not, data are a part of almost every area of our lives. As individuals, fitness trackers and blood sugar meters let us monitor our health. Online bank dashboards let us view our spending patterns and track financial goals. As members of society, we read stories of outbreaks of wildfires in California or extreme weather events and wonder if these are mere anomalies or conclusive evidence for climate change. A 2018 study claimed that even one alcoholic drink a day increased health risks,[1] and there is considerable debate about the health benefits or risks of green tea for lowering cholesterol, vitamin C for mitigating the common cold, marijuana for chronic pain, and (sadly) even childhood vaccination. But what do all these examples mean? As a popular t-shirt proclaims: "We are drowning in data, but thirsting for knowledge."[2]

These illustrations are really about understanding something systematic or the strength of evidence for a claim. How much does my blood sugar go up if I skip my morning run or eat a Krispy Kreme donut? Are there really more wildfires in California or more extreme weather events worldwide in recent years? Exactly how much does my health risk increase from drinking one or two glasses of wine a day, as others had long recommended, compared with total abstinence?

For such questions, evidence can be presented in words, numbers, or pictures, and we can try to use these to evaluate the strength of a claim or argument. The purpose of scientific research is to gather information on a topic, turn that into some standard form that we can consider as evidence, and reason to a conclusion or explanation. A graph is often the most powerful means to accomplish this because it provides a visual framework for the facts being presented. It can answer the important, though often implicit, question, "compared to what?" It can convey a sense of uncertainty of evidence for the validity of a claim. Yet it also enabled viewers to think more deeply about the question raised and challenge the conclusion. A diagram can provide a visual answer to a problem and graphic displays can communicate and persuade.

As we illustrate in this book, graphs and diagrams have often played an important role in understanding complex phenomena and discovery of laws and explanations. To truly understand the impact of a visual framework, we must not only look at contemporary examples, we must also learn how it changed science and society. We must learn history.

A Long History

This book recounts a long history, a broad overview of how, where, and why the methods of data visualization, so common today, were conceived and developed. You can think of it as a guided tour of this history, focusing on social and scientific questions and a developing language of graphics that provided insights, for both discovery and communication.

This book has a long personal history as well. It began in October 1962, when we met as undergraduates at Rensselaer Polytechnic Institute. Sequentially we became math majors, house mates, and friends. We then did our graduate work at the same university (Princeton), both supported by Educational Testing Service's Psychometric Fellowship. There we came into contact with John Tukey, Princeton's widely celebrated polymath, who was in the process of revolutionizing the field of statistics with the idea that the purpose of data analysis was insight, not just numbers,[3] and that insight—seeing the unexpected—more often came from drawing pictures than from proving theorems or deriving equations.

Tukey's guidance proved important and prophetic as we found that whatever substantive topic we worked on, our ability to understand and communicate the evidence we gathered almost always involved viewing the data in some graphic format. Our research led us both to gravitate toward aspects of the use and development of data visualization methods. This interest spanned their applications in scientific exploration, explanation, communication, and reasoning, as well as the creation of new methods for illuminating problems so that they can be understood better.

Remarkably, for both of us, our studies of graphical methods took us back ceaselessly into the past for a deeper and more thorough understanding. Much of what seemed commonplace today turned out to have deep historical roots.

There is also a long history of research, collaboration, and writing that informed this book and prompted this account. One initial foray was the 1976 National Science Foundation *Graphic Social Reporting Project* directed by Wainer.

One of the project's tasks was to assemble a coherent group of international scholars who worked on the use of graphics to communicate quantitative phenomena and create a social network to facilitate the sharing of information. This led to several conferences, a fair number of scholarly articles (e.g., Beniger & Robyn's 1978 history of graphics,[4] and the English translation of Bertin's iconic *Semiologie Graphique* [1973]).[5] Once republished in English, Bertin's ideas spread more broadly and became useful for the work of many other scholars, most importantly, Edward Tufte's transformative books.[6] Data visualization, as a field of study, was off to the races.

A second key event was Friendly's Milestones Project.[7] It has been substantially revised and now appears at http://www.datavis.ca/milestones/, which began in the mid-1990s. At that time, previous historical accounts of the events, ideas, and techniques that relate to modern data visualization were fragmented and scattered over many fields.[8] The Milestones Project began simply as an attempt to collate these diverse contributions into a single, comprehensive listing, organized chronologically, that contained representative images, references to original sources, and links to further discussion—a source for "one-stop shopping" on the history of data visualization. It now consists of an interactive, zoomable timeline of nearly 300 significant

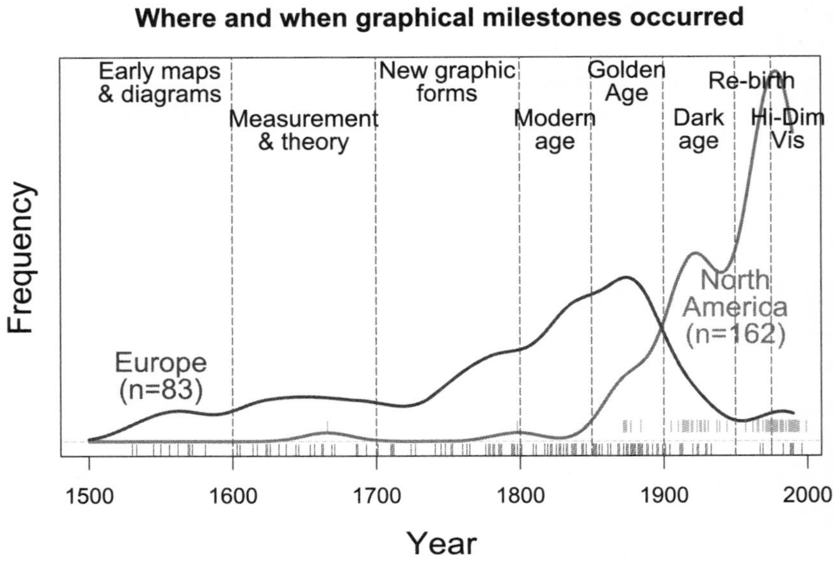

I.1 **Time line of milestone events:** Classified by place of development. Tick marks at the bottom show individual events. The smoothed curves plot their relative frequency, in Europe and North America. *Source:* © The Authors.

milestone events, nearly 400 images, and 350 references to original sources, together with a Google map of authors and a milestones calendar of births, deaths, and important events in this history.

A happy, but unanticipated, consequence of organizing this history in a database was the idea that statistical and graphical methods could be used to explore, study, and describe historical issues and questions in the history of data visualization itself. This approach can be called *statistical historiography*.[9] Each item in the milestones database is tagged by date, location, and content attributes (subject area, form of the development), so it is possible to treat this history as data.[10]

For example, Figure I.1 shows the frequency distribution of 245 milestone events classified by continent. We can immediately see that most early innovations occurred in Europe, while most after 1900 occurred in North America. The bumps in the curves reflect some global historical trends that deserve explanation. The labeled time periods provide a framework of what we consider to be the major themes driving advances in data visualization.

Overview

The earliest event recorded in the Milestones Project is an 8,000-year-old map of the town of Catalhöyük, near the present Turkish city of Konya. The prehistory of visualization goes back even further. But, as you can see in Figure I.1, most of the key innovations occurred only in the last 400 years and showed exponential growth in the last 100 years.

Our central questions in this book are "How did the graphic depiction of numbers arise?" and more importantly, "Why?" What led to the key innovations in graphs and diagrams that are commonplace today? What were the circumstances or scientific problems that made visual depiction more useful than mere words and numbers? Finally, how did these graphic inventions make a difference in comprehending natural and social phenomena and communicating that understanding?

Looking over the history portrayed in the Milestones Project, it became clear that most of these key innovations occurred in connection with important scientific and social problems: How can a mariner accurately navigate at sea? How can we understand the prevalence of crime or poverty in relation to possible causal factors such as literacy? How well are passengers and goods transported on our railways and canals, and where do we need more capacity? These are among the questions that illustrate the descriptive labels we apply to the time periods in Figure I.1.

But the story of the rise of data visualization is richer than the stimulating problems. Questions like these provide the context and motivation for many graphic inventions in this history, but they don't fully answer the question "Why?" Principal innovations over the last 400 years arose in conjunction with a cognitive revolution we call "visual thinking," the idea that some problems and their solutions can be much more clearly addressed and communicated in visual displays, rather than just words or tables of numbers. Einstein, who was better known for theories of physics expressed in words and equations, captured this visual sense in his statement, "If I can't picture it, I can't understand it."

The history we relate here is exemplified in the stories of some key problems in the history of science and graphic communication, but told as an appreciation of some of the heroes in this history, for whom visual insight proved crucial. But this begs the larger question of how such visual thinking itself developed. We provide some context for this in the initial chapters, but the essential idea is that this was bound to a concomitant rise in "empirical thinking"—the view that many scientific questions could better be addressed by gathering relevant data than by applying even the best abstract or theoretical thinking.

Re-Visions

The historical graphs we describe in this book were created using the data, methods, technology, and understanding that were current at the time. We can sometimes come to a better understanding of the intellectual, scientific, and graphical questions by attempting a reanalysis from a modern perspective.

Sometimes we come up sadly short because the software tools we have today don't allow us or make it very difficult for us to reproduce the essential ideas or the artistic beauty of important historical graphs and their stories. The hand-crafted graphs, thematic maps, and statistical diagrams of our heroes in this history often show that the pen is mightier than any software sword.

Our conscientious best efforts sometimes yield only a pale imitation of an original; in other words, we are unable to advance the understanding of the problem through reanalysis or the redrafting of graphs. One consequence is that we learn to admire the thoughtful and skillful work of our predecessors and the challenges of pen-and-ink drawings or copperplate engravings. Another consequence is that we can learn to appreciate the context of historical problems and the graphs created to present them, from both our modern successes and our failures.

We refer to these attempts as Re-Visions, meaning "to see again," possibly from a new perspective. We don't intend merely to try to see the past through present-colored glasses. Rather, we hope to shed some light on the strengths and weaknesses of the landmark developments in data visualization or understand them better in historical context. One small example illustrates this point: In Chapter 4 we show how John Snow could have made a more compelling graphic argument for cholera as a water-borne disease originating at the Broad Street pump.

Chronology versus Theme

The structure of this book requires a little explanation. In most nonfiction narratives there is considerable tension between chronology and theme, with chronology typically winning. The chronological narrative wants to move linearly from moment to moment, whereas topics scattered across eras sometimes cry out to be collected together by theme. Nevertheless chronology usually dominates, and has done so at least since narratives were recorded on papyrus scrolls.

In this book chronology dominates, but we tried to hold its force in check, fearing that if we didn't, the reader would be thematically left at sea, with the next instance far off on some foreign shore. The great themes of epistemology, scientific discovery, social reform, technology, and visual perception move with time, but not in lockstep. Consequently, much of our narrative is structured around key problems of a given time and the individuals—our graphic heroes—whose visual insight and innovations led to advances in data visualization and science.

What follows is a synopsis of the book.

Chapter 1, "In the Beginning . . . ," is an overview of the larger questions and themes that provide a context for the book. We consider the relations among numerical data and evidence for an argument and graphs, and then describe some of the prehistory of the visual representation of numbers and the early rise of visualization itself. The story continues to the rise of empirical thinking in philosophy and science around the sixteenth century and the concomitant remarkable development of the visual representation of numbers to communicate quantitative phenomena.

From there we explore a fundamental and difficult problem of the seventeenth century: the determination of longitude at sea. In Chapter 2, "The First Graph Got It Right," we show how Michael Florent van Langren had the idea to make a graph of historical determinations of the longitude distance from Toledo to Rome, in what is arguably the first graph of statistical data.

In Chapter 3, "The Birth of Data," we trace the role of data in the initial rise of graphical methods around the early 1800s. We focus attention on one important participant in this story: André-Michel Guerry [1802–1866], who used an "avalanche of data" and graphical methods to help invent modern social science.

A short time later, analogous widespread data collection began in the United Kingdom, but this was in the context of social welfare, poverty, public health, and sanitation. In Chapter 4, "Vital Statistics," we see two new heroes of data visualization, William Farr and John Snow, who worked independently trying to understand the causes of several epidemics of cholera and how the disease could be mitigated.

Chapter 5, "The Big Bang," details how, at the beginning of the nineteenth century, nearly all the modern forms of data graphics—the pie chart, the line graph of a time series, and the bar chart—were invented. These key developments were all due to a wily Scot named William Playfair. He can rightly be called the father of modern graphical methods, and it is only a slight stretch to consider his contributions to be the Big Bang of data graphics.

Among all the modern forms of statistical graphics, the scatterplot may be considered the most versatile and generally useful invention in the entire history of statistical graphics. It is also notable because William Playfair didn't invent it. Chapter 6, "The Origin and Development of the Scatterplot," considers why Playfair was unable to think about such things, and it traces the invention of the scatterplot to the eminent astronomer John F. W. Herschel. Scatterplots achieved great importance in the work of Francis Galton [1822–1911] on the heritability of traits. Galton's work, visualized through statistical diagrams, became the source of the statistical ideas of correlation and regression and thus most of modern statistical methods.

In the latter half of the nineteenth century, enthusiasm for graphical methods matured and a variety of developments in statistics, data collection and technology combined to produce a "perfect storm" for data graphics. The result was a qualitatively distinct period that produced works of unparalleled beauty and scope, the likes of which would be hard to duplicate today. In Chapter 7 we argue, as the chapter title implies, that this period deserves to be recognized as the "Golden Age of Statistical Graphics."

Chapter 8, "Escaping Flatland," discusses the challenges of creating displays of data. Displays are necessarily produced on a two-dimensional surface—paper or screen. Yet these are often misleading at worst or incomplete at best. The representation of multidimensional phenomena on a two-dimensional surface was, and remains, the greatest challenge of graphics. In this chapter we discuss and illustrate some of the approaches that were used to communicate multidimensional phenomena within the existing limitations.

Chapter 9, "Visualizing Time and Space," explores two general topics in the recent history of data visualization. First, graphical methods have become increasingly dynamic and interactive, capable of showing changes over time by animation and going beyond a static image to one that a viewer can directly manipulate, zoom, or query. Second, the escape from flatland has continued, with a variety of new approaches to understanding data in ever higher dimensions.

Graphs are justly celebrated for their ability to accurately present phenomena in a compact way while simultaneously providing their context. If this were all that they did, their place in scientific history would be secure. But with suitable data and the right design, they can also convey emotion. Indeed, in some instances graphs provide an emotional impact that can be likened to that of poetry. In Chapter 10, "Graphs as Poetry," we imagine a collaboration between the civil rights activist W. E. B. DuBois and the canonized graphic designer C. J. Minard to depict the Great Migration of 6 million African-Americans fleeing the racism and terror in the post-Confederacy South to the industrial North. The result of this *gedanken* collaboration provides a vivid example of how we can profit from studying the past to help solve the problems of the future. A final section, "Learning More," lists additional resources for those who wish to explore a topic in greater depth.

This print edition necessarily omits some materials that enrich our stories but fell to the cutting-room floor. Moreover, publishing constraints limited the number of color images. To partially compensate, we created an associated web site, http://HistDataVis.datavis.ca, containing all images in color, some of our more extended discussion, and biographical notes on some of our dramatis personae in this history. A happy consequence is that we can continue to keep this topic active with additional essays on related topics.

Thus, this book invites you to consider the history of data visualization from a larger perspective: a journey that began with the earliest visual inscriptions and progressed to social and scientific problems that could be understood in graphs and diagrams. Along this path, many innovations were forgotten or underappreciated, as Harry Truman noted in the opening quote. The following chapters highlight contributions that are imperative to the history of visual thinking and graphic communication.

1

In the Beginning . . .

If you would understand anything, observe its beginning and its development.

—ARISTOTLE, *Metaphysics*

This book invites you to think about the history of graphic communication and modern data visualization. Visual displays—graphs and diagrams—are ubiquitous today. We see them daily in popular media as weather charts, diagrams designed to explain stories about the economy, and political elections or to show topics that are trending on Twitter. In scientific articles and presentations, graphs and diagrams are widely used to convey to the eye a simple description of a discovery, a conclusion, or some process or algorithm that supports a scientific argument. In applied science, researchers now commonly use graphical methods to explore complex data and highlight an important signal against a background of noise.

Given what we see today, it is hard to imagine that this wasn't always so—that graphic displays weren't commonplace and easily understood as a visual portrait of some numeric facts. To appreciate this history, it is helpful to start at the beginning.

Throughout history, ideas and phenomena have been expressed in three different forms—words, numbers, and pictures. Words, which were initially just verbal utterances for names of things and actions, developed in early human species about 100,000 years ago. The key feature was that spoken, audible sounds came to be recognized as signifying something concrete that could be understood by others. We can imagine, perhaps fancifully, that "berries here," *"Danger—a LION!"* and even "I want you for my mate" were among the early meanings conveyed by human speech. But spoken words are ephemeral; they leave a trace only in the mind of the listener, and then only for a short time.

It was only much later that words for objects and concepts found a physical expression in inscriptions, initially as pictographs on clay tablets in Mesopotamia around 3100 BCE and independently in Egypt, China, and elsewhere. These were picture-writing systems: a sequence of pictographs,[1] such as Egyptian hieroglyphs, could be used to tell a story of a conquest or the life of a pharaoh or even record the mundane facts of a harvest or a debt.

Written language originated in pictures. Like later alphabetic writing systems, the key feature was that they were productive: A limited number of iconic symbols could be used to express a nearly infinite number of thoughts and ideas. However, the first physical inscriptions did not appear as a sign by the watering hole warning of a lion or in a marriage proposal; rather, they were used to record numbers.

The idea of a number is very old indeed, and ancient ways of writing down numbers can be traced back to Paleolithic tally sticks (dating from the Aurignacian, approximately 30,000 years ago), in which notches were cut into a bone, ostensibly to represent counts of something of interest, such as to keep track of domestic animals. A notch was made when an animal was released to pasture; later, when it returned, the shepherd's thumb would be moved down the stick, notch by notch. If the last returning animal ended with the last notch, the shepherd would be assured that all was well. Such a system, while a vast improvement on trusting counts to memory, had room for improvement. Adding an additional animal (a birth) was easy—just add another notch—but subtracting (due to a predator in the neighborhood or mutton for lunch) was more difficult and might require carving a new stick. Keeping separate track of different kinds of animals (say, goats and sheep) could be done with additional sticks for each type, but it would quickly become cumbersome to carry them around and to remember which was which.

Over time, the system of counting sticks evolved, as shown in Figure 1.1. Around 3300 BCE the Sumerians in Mesopotamia used cuneiform symbols on clay tablets to record information about trade and agriculture (Figure 1.1a). The pre-Columbian Incans in South America transformed counting sticks into *quipus*, which were ropes into which were tied knots that served the same role as notches. But knots could be untied, allowing subtraction (Figure 1.1b). Also, two or more such ropes, each representing a different kind of animal, could be tied together and easily carried wrapped around the waist or thrown over the shoulder.

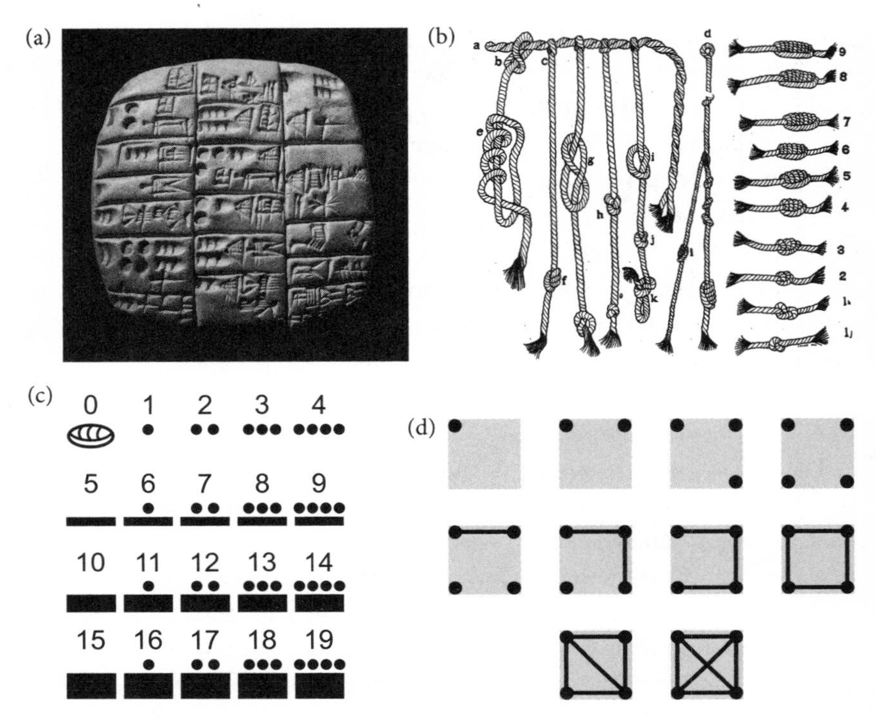

1.1 **Some of the graphic forms used to represent numbers:** (a) A clay cuneiform tablet dated as 3300–3100 BCE, giving an account of yields of barley; (b) *quipus*, a system of knots tied in ropes used by South American Incas, around 1000 CE; (c) symbols used in Mayan culture, around 500 CE, showing the numbers 0–19; (d) a scheme proposed by John W. Tukey (1977) to tally counts of observations by hand, using dots and lines in groups of 10. *Sources:* (a) Britannica.com; (b) L. Leland Locke, *The Ancient Quipu*, Washington, DC: The American Museum of Natural History, 1923, fig. 1; (c) Neuromancer2K4 / Bryan Derksen / Wikimedia Commons / GNU Free Documentation License; (d) Pinethicket / Wikimedia Commons / GNU Free Documentation License.

The Mayan culture around 100 BCE used a sophisticated number system, probably the most advanced in the world at the time (Figure 1.1c). It used base 20, most likely developed from counting on fingers and toes. The numerals from 0 to 19 used only three symbols, representing zero (a shell shape), one (a dot), and five (a horizontal bar). After the number 19, larger numbers were written in a vertical place value format using powers of 20: 1, 20, 400, 8000, and so on; thus, the number 826 was represented by symbols for $(2 \times 400) + (1 \times 20) + 6$. This made it relatively easy to both add and subtract.

The Mayans made extremely accurate astronomical observations and measured the length of the solar year to a high degree of accuracy—their calculations produced 365.242 days, compared to the modern value of 365.242198. Counting frames, which later evolved into the abacus, also arose in various ancient cultures; some of these, like the Chinese *suanpan* (2000 BCE), allowed multiplication and division. *Suanpan* arithmetic was still taught in China until handheld calculators became available in the 1980s.

Each of these early forms for representing numbers had both a physical and a visual representation. More importantly, some of these, like *quipus* and Mayan symbols, provided rudimentary ways to think about, compute with, or display quantitative facts. How many more or fewer sheep are there this year compared to last year? How many days have there been since the last full moon?

However, numbers shown in tables have a visual form that makes it hard to see patterns; for example, you have to read the numbers 826, 765, 919 to understand that the middle one is the smallest. Roberto Bachi[2] developed a notation called "graphical rational patterns," a scheme that represents numbers with dots and squares whose visual weight reflects the number itself. Similarly, for tallying counts by hand, John Tukey[3] suggested a tallying scheme using dots and lines in groups of ten (Figure 1.1d) rather than the usual method of counting in groups of five.

/   //   ///   ////   ~~////~~

It is far too easy to make an error, for example, ~~/////~~, when you meant five.

The Evolution of Pictures

Perhaps the best known of the very early examples of human visualizations are found in the Lascaux caves near the French village of Montignac in the Dorgogne region. The cave walls contain nearly 2,000 remarkable drawings of animals, human-like figures, and abstract or geometric signs, which carbon dating has estimated to be about 17,300 years old. The inhabitants of these caves were among the oldest known modern humans (*Homo sapiens*), called Cro-Magnon man.

A small section of what is now called the Chamber of the Bulls is shown in Plate 1. It is hard to get a sense of their majesty from this one image, but the collection is spectacular by any measure. Henri Édouard Prosper Breuil[4] [1877–1961], the first archaeological expert to view the cave, called the drawings in the Axial Gallery of the cave "The Sistine Chapel of prehistoric man." After seeing them in the public re-creation of Lascaux II, a modern viewer would find it difficult to ever again think of these ancestors as primitive people.

It may seem surprising for us to start this tour of the history of visualization so early, and with images, that, although impressive artistically, might be considered too far afield. Yet, there are deeper lessons here. When we view images from this history, our questions usually are: What were they thinking? Why did they draw them? What can we learn from them today? Indeed, such questions form a main theme of this book.

We can only speculate on these early cave paintings. A naive view suggests that they might reflect a celebratory display of past hunting success. But this is wrong—other evidence shows that these early dwellers in the Vézère valley hunted reindeer primarily, and there are no images of reindeer on the cave walls.

What is clear is that they had some symbolic meaning: these early artists, having neither a written language nor number system, used a visual language to communicate a story or myths of their culture to any who viewed it. Among other features, they were able to incorporate a sense of perspective, motion, and even animation in a sequence of images. This is more than enough to ask of the legacy of these early Cro-Magnon cave dwellers in southwestern France. These early human artists painted from images in their mind's eye; they had the same inner vision, the gleam, that millennia later yielded modern data visualization. More generally, this is evidence for a cognitive revolution:[5] an inner mental world of imagery, new ways of thinking and communicating, and the emergence of the modern human mind that occurred in these early *Homo sapiens* starting around 40,000 years ago.

Going forward in time, the next development in visualizations were in what we would now call diagrams and early maps—pictorial, yet abstract, representations of information. Michael Rappenglueck, a German researcher, claimed[6] that the dot patterns on the cave walls of Lascaux represented star maps of the night sky, showing the three prominent stars of the Summer

Triangle—Vega, Deneb, and Altair—and the star cluster of the Pleiades. What is clear is that the images, if they were maps of any sort, were of the heavens, not the Earth. These early human ancestors could see the night sky, but the Earth, beyond the next few hills and valleys, was largely invisible.

Among the earliest recognizable geographical maps is a remarkably modern-looking city plan—believed to be the world's first—from Catalhöyük, near the present Konya, Turkey, dating from 6200 BCE. It was inscribed on a wall in what is considered a shrine or holy room of some sort, so its purpose was more symbolic than for wayfinding or as a representation of geographical knowledge.

Maps and diagrams later acquired larger and nobler purposes—to show visually a state of knowledge, to instruct how to do something, to convey a concrete expression of the known world or how to get from where you are to where you want to be. An early diagram, dating from 2000 BCE, is shown in Figure 1.2. It is a depiction of wrestling instructions found on the walls of the tomb of Baqt in Beni Hasan, an ancient Egyptian cemetery near the modern-day Minya. It tells a graphic story of many ways to throw your opponent to the ground, animated in a way that can compete with modern visualizations or graphic novels. We might imagine that Baqt was a wrestler as a youth and later a coach, with the diagram on his tomb serving as his final lesson.

A next step in the birth of pictures to convey useful and usable information is exemplified by early Greek maps that attempted to show the scope of the known world. The epic poems of the *Iliad* and the *Odyssey*, attributed to Homer, are the oldest extant works in western literature. They are about the Mycenaean Greeks, who flourished from about 1600–1100 BC. They tell the stories of the siege of Troy and the Greek hero Odysseus,[7] king of Ithaca, and his journey home after the fall of Troy. But they are also stories—only in words—of an emerging knowledge of geography, of places and their characteristics in the world known to the ancient Greeks, encompassed by the Aegean and Ionian Seas.

Knowledge of that world became increasingly important for exploration, trade, and conquest or defense. It was captured in the first world map (Figure 1.3) by Anaximander [c. 610–546 BCE], a Greek philosopher who lived in Miletius, a city of Ionia in present-day Turkey. His map was soon improved with the addition of more detail by the geographer Hecataeus, also of Miletius. Anaximander and Hecataeus show the known inhabited world

1.2 **Wrestling instructions:** Successive moments of a wrestling match between two people. From a drawing on the east wall of the tomb of Baqt at the Beni Hasan cemetery (ca. 2000 BCE). *Source:* Wikimedia Commons.

(the *oikoumenè*) as a tripartite circle centered on the Aegean Sea and surrounded by ocean. The design may reflect the Greek ideas of harmony and symmetry as much as real geography, but it gets the general framework of relative positions right.

The key innovation was that such early maps were concrete, visual representations of the scope of the known world. Merchants could use them to plan for selling figs and buying olives in another region. Kings could use them to think about expanding their influence or protecting the territory they had. Place names marked cities and graphic symbols showed features such as rivers, mountains, and oases. To a rough degree, one could measure distance on a map and estimate how long it would take to get from point A to point B. The map became a tool for thinking and planning.

The idea of a geographic coordinate system of latitude and longitude had to wait a few centuries for Eratosthenes [c. 276–195 BCE]. But a more accurate

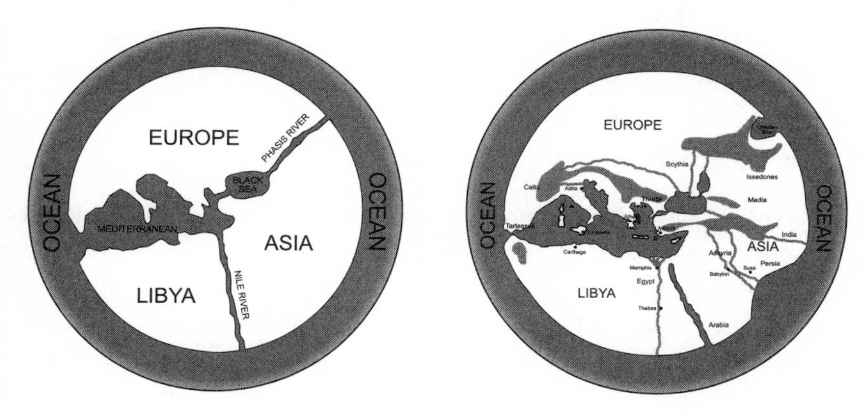

1.3 **Reconstructions of ancient Greek world maps:** Left: the world according to Anaximander of Miletus; right: a more detailed version due to Hecataeus. *Source:* Bibi Saint-Paul / Wikimedia.

map required better data. One important step came from Hipparchus of Nicaea [c. 190–120 BCE], perhaps the greatest ancient astronomer and geographer, who developed a system by determining the latitude and longitude from the heavens: latitude by using stellar measurements rather than solar altitude and longitude by the timings of lunar eclipses. Around 150 BCE, Claudius Ptolemy produced his *Geography*, a compilation of latitude and longitude of all known locations spanning 180° of longitude from the present Canary Islands in the Atlantic Ocean to the middle of China, and about 80° of latitude from the north of Scotland to the middle of Africa.[8]

Yet the idea of a coordinate system remained firmly bound to geography and maps until the seventeenth century when René Descartes [1596–1650] revolutionized mathematics as the link between Euclidean geometry and algebra by representing points, lines, and geometric figures by equations that could be visualized in diagrams and used to solve problems. In his analytic geometry, a line could be represented as the equation $ax + by = c$, and this could be graphed on (x, y) coordinate axes; a circle had the equation $x^2 + y^2 = c^2$, which also could be depicted in the diagram. Do the line and circle intersect? If so, where? This problem had a solution in algebra, but the result was conveyed directly to the eyes in the diagram. What is the distance between two points? Pythagoras had long before shown a mathematical answer in terms of the length of the hypotenuse of a right triangle; now it could be measured on the diagram with calipers. This idea of an abstract coordinate system was another crucial step in the development of visual thinking.

Connecting Data with Pictures

As we've seen, initial visualizations were of something concrete and specific in the world: a majestic auroch bull in motion, diagrams of wrestling moves, and maps of just a city or of the entire known world. But another branch of visualization was developing too, which depicted an abstract and theoretical world. A century before Descartes formalized his eponymous coordinates, Nicole Oresme in Padua illustrated some of the possible laws of motion in the book *Tractus de latitudunus forarum*, the "latitude of forms" (see Figure 2.2). Galileo and Newton would later make the study of motion precise, but Oresme had the idea to consider some alternatives and show them in a graph.

What was still lacking was a connection between empirical observations—numbers—and pictures to convey them to the eye. Natural philosophy—how we learn about the world—had long had two distinct views, rationalism and empiricism, which date back at least to Plato and Aristotle. The philosophical debate has many branches, but the essential contrast was of the role of sensory experience: using observations and data in deriving knowledge, making decisions, and formulating natural laws.

Rationalists claimed that there were some innate or intuitive ideas (a point or line, the idea of language); larger ideas (a triangle or square, words for things versus words for actions) could be deduced by human intellect. For Descartes, one of the founders of seventeenth-century rationalism, the argument was captured in his famous proclamation, "I think, therefore I am." Analytic geometry was the result of mathematical reasoning applied to geometry, but Descartes also applied this approach to the mind-body problem (determining what distinguishes the corporal self, composed of matter, from the ethereal mind and soul). The laws of the universe are fixed, and they can be discovered by reason. Observations and data were useful, but they only play a supporting role.

Empiricists claimed that knowledge and natural law had to be based fundamentally on empirical evidence, not authority or abstract reasoning. The idea of a scientific method based on observation stems from Roger Bacon [1214–1292], who observed that "reasoning draws a conclusion, but does not make the conclusion certain, unless the mind discovers it by the path of experience" (Bacon, Opus Majus c. 1267; translation from Robert Burke (2002) The Opus Majus of Roger Bacon Part 2. p. 583).

There were impressive, earth-shattering discoveries in science over the next few centuries, most based on empirical observations; some were illustrated in pictures and confirmed with mathematics, but they were largely individual contributions, and not yet understood as examples of a general empiricist philosophy. The clearest instances occurred in astronomy. An observation-based revolution began with Nicolaus Copernicus [1473–1543], who formulated a theory of the solar system placing the sun at its center, to replace the Earth-centric Ptolemaic model that had held sway for over 1,500 years. Tycho Brahe [1546–1601] meticulously cataloged astronomical and planetary observations to an accuracy far surpassing anything previously available. Johannes Kepler [1571–1630] later used Brahe's data to formulate his laws of planetary motion as elliptical orbits that could account for all known observations. Then, in 1609, Galileo Galilei [1564–1642] built one of the first telescopes, and within a few short months he discovered craters on the moon, the moons of Jupiter, rings around Saturn, and dark spots on the sun's surface (sunspots). His sketches in *Sidereus Nuncius* (1610) are still considered masterworks of visual explanation.

But it was Francis Bacon [1561–1626] who reinvigorated the formal use of evidence. This idea was subsequently expanded and amplified by the British empiricists John Locke [1632–1704], George Berkeley [1685–1753], and especially David Hume [1711–1776], whose 1738 *Treatise on Human Nature* as well as his 1741 *Essays, Moral and Political*, had a profound influence on other thinkers.

By the middle of the eighteenth century the epistemological seeds sown by the British empiricists had started to bear fruit. The Scottish enlightenment, a magical period of that century, gave rise to a torrent of practical innovations in mathematics, science, and medicine. James Watt revolutionized manufacturing; Adam Smith's *The Wealth of Nations* started modern economics; and the mathematician / geologist John Playfair's advocacy of Hutton's evidence-based theories bravely yielded an estimate of the age of the Earth that was very much at odds with the 6,000-year biblical estimate. But the star of our story is not the very worthy John Playfair [1748–1819], but rather his ne'er-do-well younger brother, William [1759–1823].

Early on in his working life William Playfair was a draftsman for James Watt. He later went on to become a pamphleteer, typically focusing on political arguments based on economic data conveyed in vivid, original graphical forms. Thus evolved the union between empiricism and visualization that began long ago during the Golden Age of Greece and completed in the latter half of the eighteenth century.

Seeing the Unexpected

Graphs that were in existence before 1800, the time of William Playfair (see Chapter 5), largely grew out of the same rationalist tradition that yielded Descartes's coordinate geometry—the plotting of curves on the basis of an a priori mathematical expression (e.g., Oresme's "pipes," shown in Figure 2.2).

The plotting of real data had a remarkable, and largely unanticipated, benefit. It often forced viewers to see what they hadn't expected. The frequency with which this happened gave birth to the empirical modern approach to science which welcomes the plotting of observed data values with the goal of investigating suggestive patterns.

This was particularly true of Playfair's graphs, most of which showed mundane economic data over time: balance of trade with other countries, the national debt, and so forth. But these had never been seen before in a way that could suggest patterns, trends, and explanations. In this period, the idea of a graph of numbers, supporting an argument based on evidence, was born.

This crucial change in view of the value of graphs in relation to evidence and explanations has a more nuanced history than we can tell here. However, the revolution seems to have begun in 1665 with the invention of the barometer and graphic recording devices, which used pens driven on paper by a measuring instrument.[9] Readings from this instrument inspired the eponymous Robert Plot to record the barometric pressure in Oxford every day of 1684 and summarize his findings in a strikingly contemporary graph that he called a "History of the Weather" (Figure 1.4).

This graph is not a beautiful example of a data plot. It looks more like a recording of an old polygraph or cardiac ECG monitor—a bunch of squiggly lines on a dark gray background, with ruled lines that further obscure the data. But the visual insight he had from this, and what he could see as an eventual wider use of plots of the weather, was important. His idea of

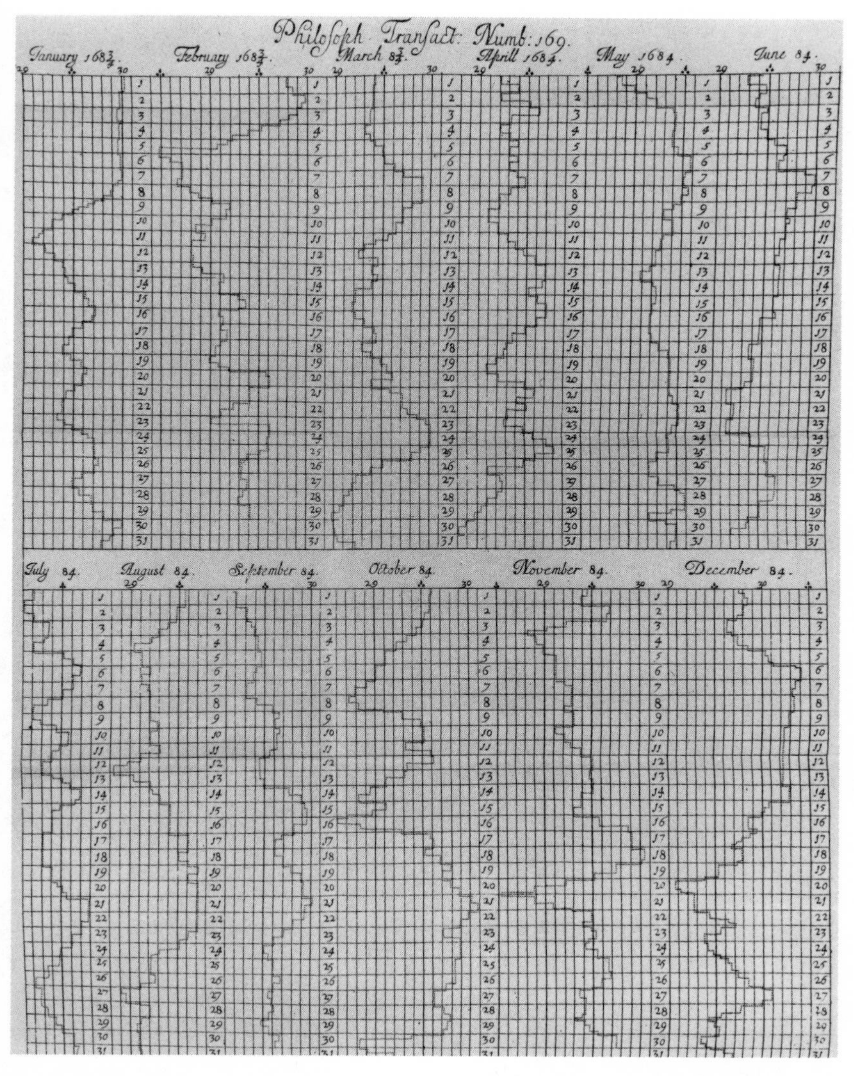

1.4 **Robert Plot's *History of the Weather:*** The graph records the daily barometric pressure in Oxford for the year 1684. *Source:* Robert Plot, "Observations of the Wind, Weather, and Height of the Mercury in the Barometer, throughout the Year 1684; Taken in the Musaeum Ashmoleanum at Oxford," *Philosophical Transactions*, Vol. 15, 930–943.

recording a history of weather made the phenomenon of barometric pressure subject to visual inspection and scientific thought.[10]

In the same year Plot sent a copy of the graph to the naturalist Martin Lister [1639–1712], a fellow of the Royal Society, with this prophetic description and a call for further data to turn his history of weather into a science of weather:[11]

> For when once we have procured fit persons enough to make the same Observations in many foreign and remote parts, how the winds stood in each, at the same time, we shall then be enabled with some grounds to examine, not only the coastings, breadth, and bounds of the winds themselves, but of the weather they bring with them; and probably in time thereby learn, to be forewarned certainly, of divers emergencies (such as heats, colds, dearths, plague, and other epidemical distempers) which are not unaccountable to us; and by their causes be instructed for prevention, or remedies we shall certainly . . . obtain more real and useful knowledge in matters in a few years, then we have yet arrived to, in many centuries.

Plot clearly recognized the power of a graph to inform, forewarn, and find remedies. His use of graphic displays of data had been anticipated, though more simply, by the Dutch polymath Christiaan Huygens [1629–1693]. In 1662, John Graunt published the first data on life expectancy in his *Natural and Political Observations on the London Bills of Mortality*. On October 30, 1669, Christiaan's brother Lodewijk sent him a letter containing some interpolations from these data, calculating life expectancy for ages not shown in Graunt's numerical table. Christiaan responded in letters dated November 21 and 28, 1669, with graphs of those interpolations.

Figure 1.5 contains one of those graphs showing age on the horizontal axis and number of survivors of the original birth cohort on the vertical axis. The actual data points are shown as labels on the curve. The curve Christiaan drew was fitted to his brother's interpolations, but they convey a more general view: as shown by the letters for lines in the chart, one could draw a vertical line for any given age and estimate the number surviving. Christiaan thought that this was most interesting from a scientific point of view; indeed, his simple use of interpolation and addition of a smoothed curve was a signal advance in the emerging connections between data and graphs and scientific and practical

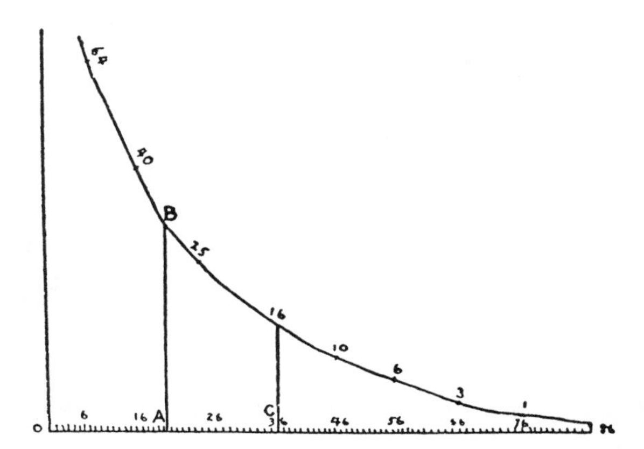

1.5 **Huygens's survival graph:** Christiaan Huygens's 1669 curve showing how many people out of a hundred survive between the ages of infancy and eighty-six. The data are taken from John Graunt's *Natural and Political Observations on the Bills of Mortality*, 1662. *Source:* Christiaan Huygens, *Oeuvres Complètes*, Volume 6. La Haye: M. Nijhoff, 1895.

uses. The graphic idea of a smoothed curve of life expectancy soon found use as a method for setting prices for life insurance and annuities.[12]

A smattering of other examples of empirically based graphs appeared in the century between Huygens's letter and the 1786 publication of Playfair's *Commercial and Political Atlas*, but these were not common. Albert Biderman[13] argued that this was due to skepticism and even antipathy toward empirical graphs as a scientific approach. At the beginning of the nineteenth century there occurred an explosive growth in the use of data graphics and we argue (Chapter 3) that this was in no small part due to the exponential growth in the availability of data, particularly in the social sciences. The empirical approach to problem solving, a critical driving force for data collection, was slow to get started. But this began to demonstrate unprecedented success in understanding and solving problems; with improved communications, the news of these successes, and hence the popularity of the associated graphic tools, began to spread quickly.

We are accustomed to intellectual diffusion taking place from the natural and physical sciences into the social sciences; certainly that is the direction taken for both calculus and the scientific method. But statistical graphics in particular, and statistics in general, took the reverse route. Although, as we have seen, there were applications of data-based graphics in the natural sciences, it was only after Playfair applied them to economic data that their popularity began to accelerate. Playfair should be credited with producing the first chartbook of social statistics. That he published an atlas that contained not a single map is one indication of his belief in the methodology (and his audacity). Playfair's work was immediately admired, but emulation,[14] at least in Britain, took a little longer (graphic use started up on the continent a bit sooner).

The relatively slower diffusion of the graphical method back into the natural sciences provides additional support for the reluctance toward empiricism there. The newer social sciences, having no such tradition and faced with both problems to solve and relevant data, were quicker to see the potential of Playfair's methods.

The Rise of the Graphic Method and Visual Thinking

Playfair's graphical inventions—the line chart, bar chart, and pie chart—are the most commonly used graphical forms today. The bar chart was something of an anomaly: lacking the time series data required to draw a timeline showing the trade with Scotland, he used bars to symbolize the cross-sectional character of the data that he did have. Playfair acknowledged Priestley's (1765, 1769) priority in this form, using thin horizontal bars to symbolize the life spans of historical figures in a time line (Figure 1.6). What attracted Playfair's interest was the possibility of visualizing a history over a long period and showing a classification (statesmen versus men of learning)—all in a single view.

Playfair's role was crucial for several reasons. It was not for his development of the graphic recording of data; others preceded him in that. Indeed, in 1805 he pointed out that as a child his brother John had him keep a graphic record of temperature readings. But Playfair was in a remarkable position. Because of his close relationship with his brother and his connections with Watt, he was on the periphery of science. He was close enough to know of the value of the graphical method but sufficiently detached in his own interests to apply them in a very different arena—that of economics and finance. These areas, then as now, tend to attract a larger audience than matters of science, and Playfair was adept at self-promotion.[15]

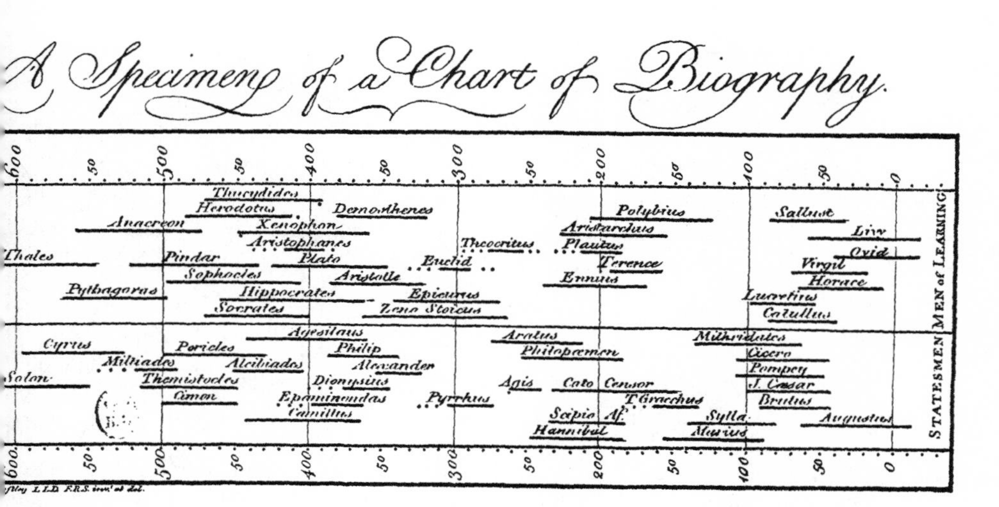

1.6 **Chart of biography:** Lifespans of fifty-nine famous people in the six centuries before Christ. *Source:* Joseph Priestley, *A Chart of Biography*, London, 1765.

In a review of his 1786 *Atlas*, which appeared in *The Political Herald*, the Scottish historian Dr. Gilbert Stuart wrote,

> The new method, in which accounts are stated in this work, has attracted very general notice. The propriety and expediency of all men, who have any interest in the nation, being acquainted with the general outlines, and the great facts relating to our commerce are unquestionable; and this is the most commodious, as well as accurate mode of effecting this object, that has hitherto been thought of. . . . To each of his charts the author has added observations (which) . . . in general are just and shrewd; and sometimes profound. . . . Very considerable applause is certainly due to his invention; as a new, distinct, and easy mode of conveying information to statesmen and merchants.

Stuart appreciated the power of a graphic history of economic data accompanied by a narrative of words. The visual form was conveyed to the eye of the viewer; the text provided explanations and conclusions that could be assessed against the evidence in the chart. Such wholehearted approval rarely greets any scientific development. Playfair's adaptation of graphic methods

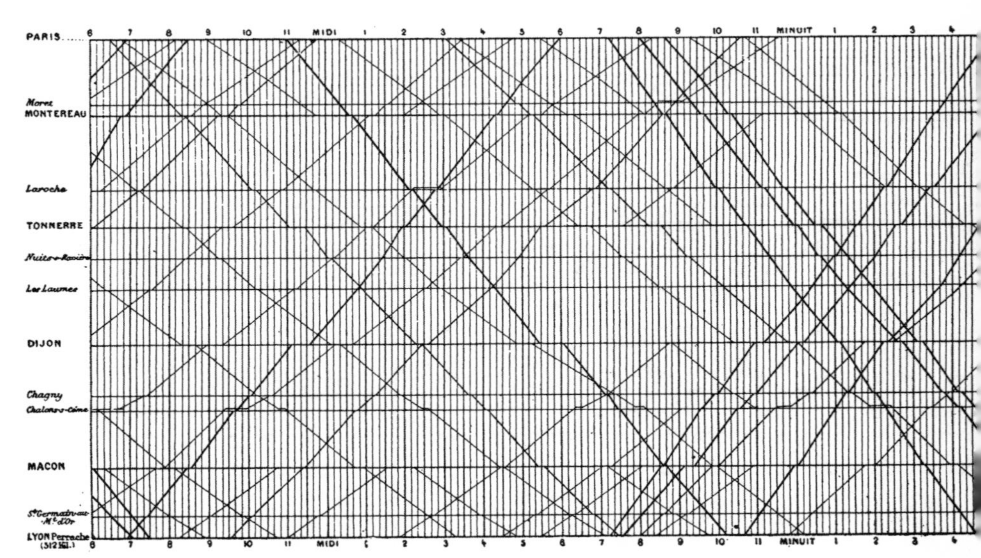

1.7 **Graphical train schedule:** E. J. Marey's (1878) graphical train schedule, showing all trains between Paris and Lyon each day. *Source:* Étienne-Jules Marey, *La méthode graphique dans les sciences expérimentales et principalement en physiologie et en médecine*. Paris: G. Masson, 1878.

to matters of general interest, to both statesmen and merchants, provided an enormous boost to the popularity of statistical graphics.

By the mid-1800s, a new view of the role of visualization in scientific discovery and explanation had been implanted. There were many participants, whose contributions form the bulk of this book. Among these, we find the French physiologist Étienne-Jules Marey [1830–1904], who in 1878 wrote *La méthode graphique*, an attempt to found a new approach to scientific questions through a direct appeal to graphs as the primary tool (see Chapter 9). A simple example is shown in Figure 1.7, a graphic schedule of all the trains between Paris and Lyon, which is also from 1878. The topic is prosaic, but the graph illuminated a new aspect of visual thinking.

Each line in the graph shows a different train as a line, from its origin to destination. The stations are spaced according to distance, so steeper lines indicate faster trains. Whether this was of any help to travelers is unknown, but it reflected Marey's deep conviction that the graphical method could be applied to almost any question. He said boldly,

> There is no doubt that graphical expression will soon replace all others whenever one has at hand a movement or change of state—in a word, any phenomenon. Born before science, language is often inappropriate to express exact measures or definite relations. (Marey, 1878, p. iii)

The graphic representation of scientific phenomena served two purposes. Its primary function was to make previously invisible phenomena subject to direct inspection in a graphic display. Marey's *The Graphic Method* was not the first or only expression of this view, but it was revolutionary in scope and vision. In this and other work, he laid out the essential ideas of ingenious devices to record blood pressure, heart rate, the flapping of wings of birds and insects, the exertions of a sprint runner, and so forth, to turn graphic recording of empirical observations into subjects for scientific study. Like Playfair, Marey was motivated by an inner vision of graphic display.

The graphic method had another function, that of communication to the scientific community and educated readers. These displays made complex phenomena palpable and concrete. They preserved what was ephemeral and distributed it to all who would read the volume, not just those on hand at the right place and the right time and with the right equipment to see them with their own eyes. They served the cause of memory, for images are more vivid and indelible than words.

Pictures became more than merely helpful tools: to Marey, they were the words of nature herself, recorded for all to see.

A Golden Age

Something even more remarkable occurred in the latter part of the nineteenth century, as many forces combined to produce the perfect storm for data graphics, in what we call the Golden Age of Graphics (Chapter 7). By the mid-1800s, vast quantities of data on important social issues (commerce, disease, literacy, crime) had become available in Europe and the United States, so much that one historian called this an "avalanche of numbers."[16] In the second half of this century some statistical theory was developed to allow their essence to be summarized and sensible comparisons to be made. Technological advances in printing and reproduction now allowed for the broad dissemination of graphic works in color and with a graphic style that was unavailable previously. Excitement and enthusiasm for graphics were in the air.[17] The audience was international, but they shared a common visual language and visual thinking.

Another of the key developers of graphic vision was Charles Joseph Minard [1781–1870], a civil engineer in France, who produced what is now applauded as the greatest data-based graphic of all time—a flow map depicting Napoleon's disastrous Russian campaign of 1812 (see Figure 10.3). Minard used the graphic method to design beautiful thematic maps and diagrams showing all manner of topics of interest to the modern French state during the dawn of national concern for trade, commerce, and transportation: Where to build railroads? How did the US Civil War affect the British mills' importation of cotton? In these and other graphs, he told graphic stories of immediate visual impact—the message hit the viewer between the eyes. Minard too was driven by an inner vision.

By the end of the nineteenth century, scientists from the United States (Francis Walker in the Census Bureau), France (Émile Cheysson in the Ministry of Public Works), and others in Germany (Herman Schwabe, August F. W. Crome), Sweden, and elsewhere, began to produce and widely disseminate elaborate and detailed statistical albums tracing and celebrating their nations' achievements and aspirations. These contain some of the most exquisite graphs ever produced, even to this day. They were resplendent in color and style and revealed a vision of inventive graphic design that serves as a model to emulate and has become part of the language of graphics today. They inspire awe, just as do the cave paintings in Lascaux.

In this chapter we have taken a long-range view of visualizations spanning more than 17,000 years from the Lascaux cave paintings of long extinct auroch bulls to Minard's equally exquisite depictions of the horrors of the Napoleonic Wars. In both these cases, and in many in between, visualizations have painlessly provided memorable understanding for those who look at them. Over the course of centuries, rising visual thinking was expressed in diagrams, maps, and graphs. A universal language of visualizations was used to communicate both quantitative and qualitative information, to uncover complex phenomena, to support, or refute, scientific claims. In the balance of this book we will elaborate and illustrate the wonders of visual communication and welcome you along on the journey.

2

The First Graph Got It Right

Who invented the idea of graphing data? This question is meant to be provocative and contentious. It is provocative because you might wonder or argue about what "the idea of graphing data" actually means: What counts as "data"? What counts as "graphing"? It is also a deliberatively contentious question, because for graphs, like many other scientific discoveries and inventions, claims of "firsts" are difficult to pin down: many important developments were preceded by other contributions that could count as the initial one under some looser, more relaxed definition; conversely, those who followed often developed the idea in wider or more general ways.[1]

As one example, the idea to record geographical positions in terms of latitude and longitude on a map is widely attributed to Claudius Ptolemy in his *Geographia* around 150 AD. An earlier description of a geographic coordinate system goes back to Eratosthenes of Cyrene in the 3rd century BCE. And among the earliest world maps, Anaximander of Miletus [c. 611–546 BCE] sketched the known world in a circular diagram showing Europe, Africa, and Asia separated by bodies of water and surrounded by (unknown) oceans (Figure 1.3). An even earlier depiction of the world known to the Babylonians was found on a clay tablet at Sippar, southern Iraq, and dated to the 5th century BCE.

Going forward in time from Ptolemy, map information became progressively more detailed and precise over time, and map makers developed new ways to project the 3D globe onto a planar map. So, the question of who invented the idea of mapping the world may not have an unambiguous answer, but the discussion of this topic can be informative.

We see a similar trajectory with the rise of statistical graphs and start with a specific example to illustrate the intellectual and scientific context within which statistical graphs first arose. We credit the idea of graphing data to the

Dutch cartographer Michael Florent van Langren in the period from 1628 to 1644.

Michael van Langren [1598–1675] was the third generation of a family of prominent Dutch cartographers, globe makers, astronomers, and mathematicians in the period from the mid-1500s to the end of the 1600s. His grandfather, Jacob Floris van Langren [ca. 1525–1610], produced the first world globes, beginning around 1586. These globes were significant in the development of Dutch seafaring trade. Michael's father, Arnold van Langren [1571–1644], learned the crafts of engraving copper plate and globe making. By all accounts, Arnold was a skilled engraver but somewhat of a ne'er-do-well. By late 1607 or early 1608, he had accumulated sufficient debts that he was forced to flee from Amsterdam to the Spanish-ruled southern provinces—in such haste to escape the bailiff that he left behind his household goods and engraving tools.

However, though he lacked business sense, his social skills included the ability to attract supporters in high places. In September 1609, Arnold succeeded in being named official Spherographer to Archduke Albert of Austria and his wife, Infanta (Princess) Isabella Clara Eugenia of Spain (daughter of Philip II), rulers of the Spanish Netherlands. Some of Arnold's work on terrestrial and celestial globes and the accompanying manuals needed for mariners at sea was carried out by Arnold's son, Michael Florent van Langren, who now found himself with a powerful potential patron.

Among van Langren's contributions were to determine longitude. His efforts to show the extent of the problem in visual form resulted in the first graph of statistical data. The best-known version, shown in Figure 2.1, is a simple one-dimensional dot plot, showing twelve estimates of the distance, in

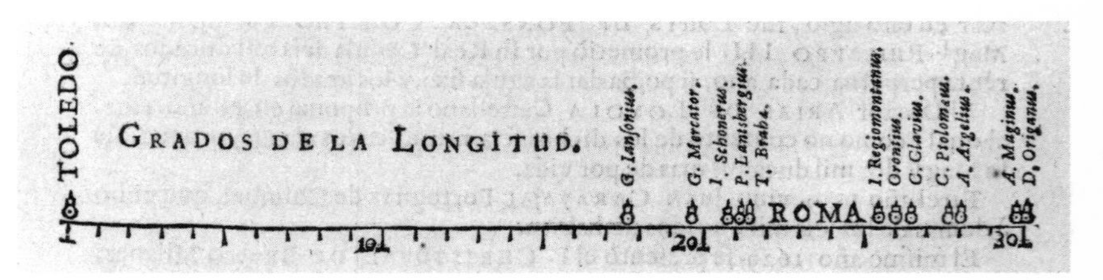

2.1 **The first statistical graph:** Van Langren's 1644 graph of twelve determinations of the longitude distance from Toledo to Rome: The correct distance is 16.5°. *Source:* M. F. van Langren, *La Verdadera Longitud por Mar y Tierra*, Antwerp, 1644. Reproduction courtesy of the Koninklijke Bibliotheek van België.

degrees longitude, between Toledo and Rome. This graph, which is conventionally dated to a 1644 publication, *La Verdadera Longitud por Mar y Tierra* (The true longitude for sea and land), is an important milestone in the history of data visualization, an image that communicated better than mere words or numbers.

Van Langren's story is of interest here because it illustrates (a) an attempt to solve a crucial problem of the times (the precise determination of longitude), (b) the role played by patronage in early scientific discovery, and (c) a cunning awareness of how a picture could communicate far better than mere words or numbers.

Early Things Called "Graphs"

To see why van Langren's graph is important, consider what had come before. The first known image of any sort that could be called a "graph" is an anonymous tenth-century conceptual depiction of cyclic movements of the seven most prominent heavenly bodies through the constellations of the zodiac described by Funkhouser (1936) and reproduced in Tufte (1983, p. 28). It is a graph because it shows curves with (x, y) coordinates, but it is more properly considered a sketch or schematic diagram because it is not based on any data.

The next step occurred around 1360, when Nicole Oresme [1323–1382], a polymath philosopher of his time, conceived of the idea to visualize how any two physical quantities (such as time, velocity, or distance traveled) might vary in a lawful, functional relation. In the *Tractatus de latitudinibus formarum* (Latitude of Forms), published only much later (Oresme, 1482), he used the terms "latitude" and "longitude" in much the same way as we now use *ordinate* and *abscissa*, anticipating Descartes (1637) in this regard by over 250 years. His diagrams, shown in Figure 2.2, are the earliest abstract graphs we know of.

In this clip from one page, he illustrates a quadratic function, a function approaching an asymptote, and a linearly decreasing function. He even anticipates the now widely deprecated 3D versions of bar charts available in Microsoft Excel and other charting software. The only thing not to like is that they also were not based on any data and so do not count as data graphs. Of this, Funkhouser said, "If a pioneering contemporary had presented Oresme with actual figures to work upon, we might have had statistical

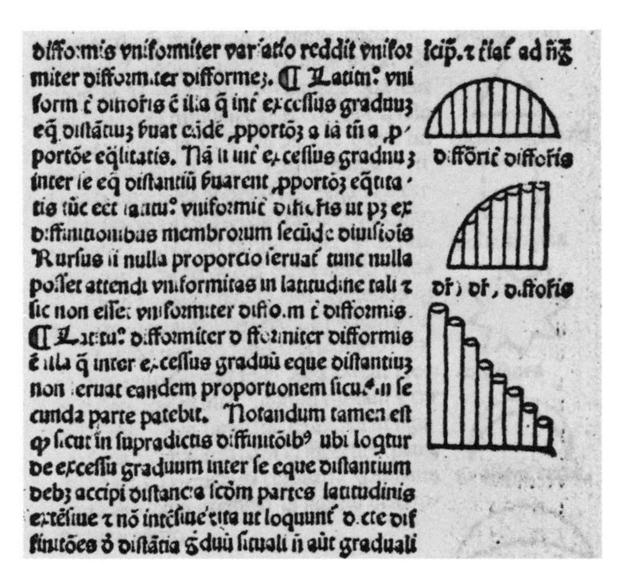

2.2 **Graphs of functional forms:** A portion of a page from Oresme's *Latitude of Forms*, showing three graphical forms arising from a functional relation between physical variables. *Source:* Nicole Oresme, *Tractatus de latitudinibus formarum*. Padua: Matthaeus Cerdonis, 1482.

graphs 400 years before Playfair."[2] But empirical data that might have been displayed graphically were relatively unknown until about 1600, when van Langren entered the scene.

To help us appreciate van Langren's contribution in historical context, Figure 2.3 shows a timeline of the major inventions of basic forms of data graphs over the period 1600–1850. Most of the well-known present-day graphical methods were invented only in the last 100 years of this timeline. Van Langren's graph, over 200 years prior, stands out as an early outlier.

The Problem of Longitude

In van Langren's era, the most important scientific questions of the sixteenth and seventeenth centuries concerned physical measurement—of time, distance, and spatial location. These arose in astronomy, surveying, and cartography and were related to practical concerns of navigation at sea, exploration, and the quest for territorial and trade expansions among European states.

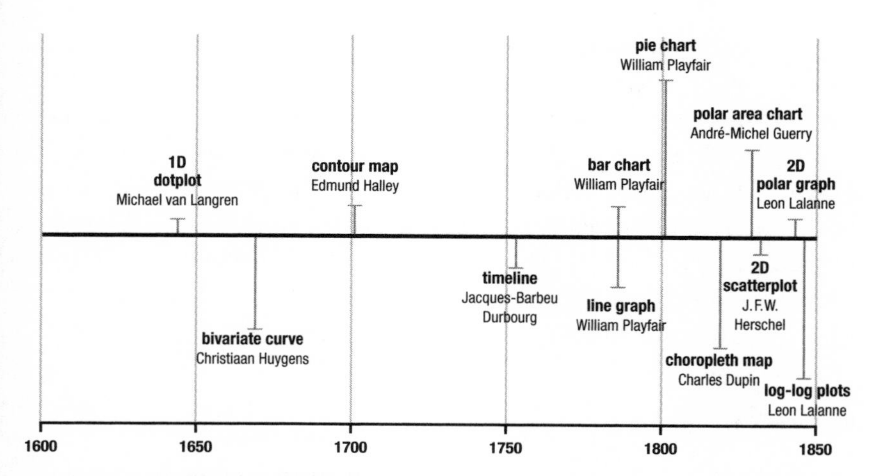

2.3 **Graphical inventions:** Timeline of the invention of some basic forms for statistical graphs, 1600–1850. *Source:* © The Authors.

Among the most vexing—and also consequential—problems was the accurate determination of longitude, on land and at sea. At best, errors in navigation led to much longer travel times and short rations or starvation for seamen; at worst, they led to numerous shipwrecks and maritime disasters. This story illustrates our theme of the connection between the visualization of some phenomenon and the explanation or solution of a well-posed problem.

Latitude, a North–South position relative to the equator, has a physical 0 point on the terrestrial sphere and a conventional range to ±90° at the poles; this can easily be found with a sextant or other device for measuring the angle (declination) of the sun, moon, or given stars and using tables of their positions that had been common for many centuries. Longitude, the East–West position, has no natural 0 point and no natural points of reference; the only physical fact is that the scale for latitude of 0–360° around any parallel corresponds to the 24-hour rotation of the Earth, or 15° per hour.

What was needed for longitude was an accurate means to determine the difference in time from where you are to some fixed reference point. For example, if a navigator knew the time in Toledo was 3 PM at the same instant the sun reached its zenith (local noon), the difference between local noon and Toledo time, 3 hours, would give the longitude difference, 45° west of Toledo. But how could a navigator know the time in Toledo out at sea? There were two classes of solutions based on clocks and books recording celestial observations.

The ultimate solution would prove to be a marine chronometer, accurate enough to keep time precisely to a few seconds over a 1-month journey. As early as 1530, Reginer Gemma-Frisius [1508–1555] gave a theoretical description of determining longitude by the difference between two clocks. But this method would remain insufficiently accurate until the self-educated English clockmaker John Harrison [1693–1776] fabricated his first Sea Clock (H1), which passed a sea trial across the Atlantic Ocean in 1736.

Show Me the Money

The longitude problem was so important that various European countries offered substantial prizes for "the discoverer of longitude." The first was offered by King Philip II of Spain in 1567. After Philip III came to the Spanish throne in 1598, a prize of 6,000 ducats[3] plus a pension of 2,000 ducats for life was offered. Other rewards were offered by Holland and France, culminating in the Longitude Prize, begun in 1714 by the British Parliament and resulting in awards totaling over 100,000£ (O'Connor and Robertson, 1997). (The immediate stimulus for this prize was one of the worst ever maritime disasters, on October 22, 1707, in which five ships commanded by Admiral Cloudsley Shovel struck the Isles of Scilly, resulting in 2,000 sailors killed.) This was the ring into which van Langren tossed his hat. Serious prize money was at stake.

In van Langren's time, a second class of solutions to the longitude problem relied on celestial observations—positions of the sun, the moon, stars, or planets, together with astronomical "ephemeris" tables giving the times of those positions at a given location. As early as 1514, Johannes Werner [1468–1522] proposed what became known as the lunar distance method: determining longitude by measuring the angle between the moon and some star or the sun, together with the use of a nautical almanac recording those positions at given dates and times at a fixed position. For instance, on van Langren's thirtieth birthday, April 27, 1628, a mariner might measure an angle between the moon and the North star, Polaris, as 30.0 degrees at midnight. He could then look up in his copy of the Alfonsine Tables (sometimes spelled Alphonsine

Tables) and learn that this event would occur at 10:30 PM in Toledo. Bingo! His longitude is 2.5 hours × 15 = 37.5° west of Toledo.

This solution to the problem was difficult enough on land, but it was far more so at sea, where celestial observations were subject to constant and changing motion, not to mention difficulties in determining local time and periods of bad weather or cloud cover. It was therefore often seriously prone to error. Over the next 200 years, problems that stemmed from attempts to solve the longitude problem by means of celestial observation and reference tables occupied the attention of the best astronomers and mathematicians of the times, from Galileo's 1612 tabulations of the orbits of the moons of Jupiter to Edmund Halley's 1683 recordings of transits of the moon across various stars. In 1750 Tobias Mayer suggested the need to take account of small perturbations ("librations") in observed lunar motion, and fifty years later, nearly simultaneously, Adrien-Marie Legendre in France and Carl Friedrich Gauss in Germany formalized the problem of combining observations subject to error in the method of least squares, which is the root of modern statistical methods. Mayer and Gauss both received small prizes from the British Longitude Commission for their work.

Van Langren's Graph

With the problem of longitude in mind, we now consider van Langren's 1644 graph, shown in Figure 2.1. The earliest version of this graph appears in a letter to the Infanta Isabella around March 1628 (discussed shortly).

Looking at this graph, what can we see? Above the horizontal axis, van Langren shows twelve estimates of the longitude distance from Toledo to Rome, on a scale of 0–30 degrees, each with a label (written vertically) for the name of the person who made that determination. The names, including Claudius Ptolemy, Gerardus Mercator, and Tycho Brahe, represent the most illustrious stars in the firmament of astronomy and geodesy to that time. The true distance is 16.5°, but that value would remain imprecisely known for over one hundred years. This graph is remarkable in the history of data visualization for several reasons.

First, it would have been easiest for van Langren to present this information to the Spanish court in tabular form, showing (Name, Year, Longitude

Table 2.1. Tables: Two of the possible tables van Langren might have used.

Sorted by Longitude				Sorted by Priority			
Longitude	**Name**	**Year**	**Where**	**Year**	**Name**	**Longitude**	**Where**
17.7	G. Iansonius	1605	Flanders	**150**	Ptolomeus, C.	27.7	Egypt
19.6	G. Mercator	1567	Flanders	**1463**	Regiomontanus, I.	25.4	Germany
20.8	I. Schonerus	1536	Germany	**1530**	Lantsbergius, P.	21.1	Belgium
21.1	P. Lantsbergius	1530	Belgium	**1536**	Schonerus, I.	20.8	Germany
21.5	T. Brahe	1578	Denmark	**1542**	Orontius	26.0	France
25.4	I. Regiomontanus	1463	Germany	**1567**	Mercator, G.	19.6	Flanders
26.0	Orontius	1542	France	**1567**	Clavius, C.	26.5	Germany
26.5	C. Clavius	1567	Germany	**1578**	Brahe, T.	21.5	Denmark
27.7	C. Ptolomeus	150	Egypt	**1582**	Maginus, A.	29.8	Italy
28.0	A. Argelius	1610	Italy	**1601**	Organus, D.	30.1	Germany
29.8	A. Maginus	1582	Italy	**1605**	Iansonius, G.	17.7	Flanders
30.1	D. Organus	1601	Germany	**1610**	Argelius, A.	28.0	Italy

distance, Where), as suggested in Table 2.1. Indeed, tables were the common form of recording observational data in that time. Moreover, a table could have been arranged to highlight authority (sorted by name), priority (sorted by year), or the range of values (sorted by longitude value).

At about the same time (1634), Henry Gellibrand, an English astronomer, introduced the idea of taking an average of multiple determinations of the same quantity.[4] Had van Langren known of and considered this, he might also have calculated some sort of measure (the midrange or average) combining the twelve separate estimates into a single "best" value and been done with it. But that was not his purpose.

Only a graph speaks directly to the eyes and shows the wide variation in the estimates; the most salient feature is that the range of values covers nearly half the length of the scale. It also appears from the graph that van Langren took as his overall summary the center of the range (anticipating Gellibrand), where there conveniently happened to be a large enough gap for him to inscribe "ROMA."

Second, van Langren's graph serves as a compelling visual example of the modern idea of systematic error, which is called *bias* in statistical estimates. In Figure 2.1, the true longitude distance is 16.5°, and the bias is the distance from that point (16.5) to "ROMA" (at 23.5), which is 7°. The size of this bias can be appreciated better by overlaying van Langren's graph on a modern map, as shown in Figure 2.4—nearly all previous estimates of longitude distance were extremely far from correct; in fact, they place Rome anywhere from the Adriatic Sea to Greece or Turkey.

Finally, van Langren's graph is also a milestone as the earliest-known exemplar of the principle of effect ordering for data display (Friendly and

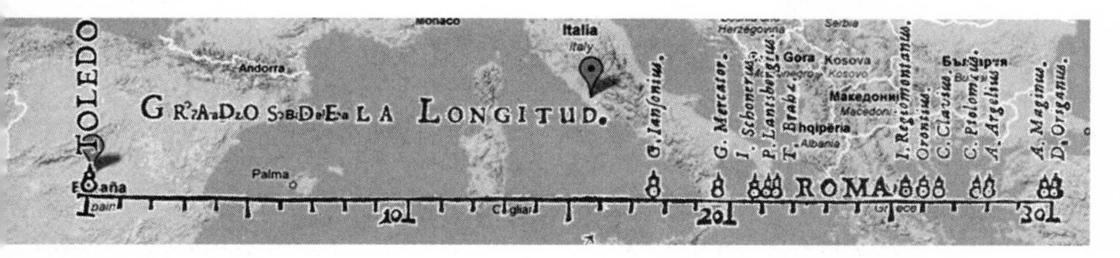

2.4 **Overlay:** Van Langren's 1644 graph, linearly rescaled and overlaid on a modern map of Europe. Toledo and Rome are shown by markers on the map. *Source:* Map: Google Maps; overlay: M. F. van Langren, *La Verdadera Longitud por Mar y Tierra*, Antwerp, 1644. Reproduction courtesy of the Koninklijke Bibliotheek van België.

Kwan, 2003): graphs and tables are most effective when the information is arranged to highlight the main features to be seen. In this case, it is clear that van Langren's main presentation goal was to show the enormous range of differences among the greatest known astronomers and geographers. As such, the graph is all the more remarkable for its focus on uncertainty or variability of observations, a topic that did not receive serious attention until roughly a hundred years later.[5] The first graph did indeed "get it right."

However, it is also fair to say that van Langren probably did not think of his image as a new species of illustration that we now call a "graph." With his family background in cartography, Figure 2.1 is essentially a one-dimensional map, where the horizontal dimension reflects a parallel of latitude and the points are possible longitude positions for Rome relative to Toledo. Indeed, John Delaney (2012), in a delightful book, *First X, Then Y, Now Z*, describes the development of thematic maps as beginning in complexity with place names ("X marks the spot"), for example, in Portolan charts that served as navigational aids for mariners traveling along a coast.

Patronage and Grantsmanship

It remains only to explain what led van Langren to contemplate putting this information in the form of a labeled dot chart rather than in tabular form, as illustrated in Table 2.1. Unlike those with independent means of support, and lacking a university education, van Langren had to depend on patronage to sustain his work and earn a livelihood. Through his father's connection to Isabella, who had become the governor of the Spanish Netherlands, he obtained commissions around 1626 to draw several maps, which he dedicated to her. Evidently, he had also learned his father's charm. Isabella soon became his patron, and by 1628 he had secured his appointment as Royal Cosmographer and Mathematician to her nephew, King Philip IV, for which he would receive an annual retainer of 1,200 écus (gold ducats, approximately 9.5 pounds of gold), certainly a considerable sum. Michael van Langren had arrived: financially and in the Spanish court.

He soon had an idea for an improvement of the lunar method to allow for more accurate determination of longitude, a goal that would occupy his attention over the rest of his life. But how could he secure credit for this contribution, as well as the attendant financial rewards? His first attempt was a letter to Isabella dated around March 1628, containing the first version of the graph

(Figure 2.1) that would later appear in *La Verdadera*. In flowery language, he makes the following points:

- *Credentials*: I am Mathematician to His Majesty, and my grandfather and father were the first to invent globes for the purpose of navigation.
- *Problem*: I have studied the most important problem—the determination of Longitude.
- *Demonstration*: You can plainly see in my chart that even the Longitude between Toledo and Rome is subject to large errors. "If the Longitude between Toledo and Rome is not known with certainty, consider Your Highness, what it will be for the Western and Oriental Indies, that in comparison the former distance is almost nothing."
- *Supplication*: Therefore, grant me a patent to solve this problem, "ordering in it that all interested in the art observe what the supplicant advises them, promising that many benefits will derive for navigation, and eternal memory for His Majesty and Your Highness, for having ordered this general correspondence of the art, and Your Most Serene Highness will receive it very particularly."

Thus, he makes explicit that the purpose of drawing a graph is to show the "countless errors" in the determination of longitude distance between two relatively well-known locations. In statistical language, his presentation goal was to show uncertainty or variability rather than a best estimate obtained from pooling the data points.

In its intent, the letter can be read as a classic example of the syllogism of a patronage request. Viewed as a modern grant proposal, what is notably missing from this letter is a Method section indicating how he intends to solve the longitude problem. Perhaps fearing that others would steal his thunder or claim priority for his method, he intentionally kept the details secret.

Eyes on the Prize

Although van Langren did not succeed in his first "grant application," he persisted over the next decade, writing further letters seeking support from various ministers in the council of Philip IV. Each of these contains another version of his graph, the statement that it reveals the "countless errors" in longitude, and the claim that he, mathematician to His Majesty, has uncovered some "very important secrets" with regard to the calculation of longitude, both on land and at sea. His letter of 1633 concludes with the "big ask":

> And in this regard His Majesty offered to the Inventor of such solution great rewards, and in particular to Luis Fonseca 6,000 ducats every year, and then to Juan Arias, 2,000 ducats every year for a lifetime; so if his Majesty sends to this Supplicant the assurance of a prize that his Royal Highness judges appropriate, he (the Supplicant) will report the aforementioned secret to His Majesty, because finding this invention and not getting any reward would be honorless.

Van Langren was moderately successful this time, for he received some unknown but handsome compensation, even without revealing his "secret."

In the 1633 letter, van Langren was sufficiently concerned for his claim of priority for his method that he states, "[My person] also supplicates that His Majesty shelters him against the objections that some could put, saying that my invention is old and known." In 1644, he decided to publish, and the result was *La Verdadera Longitud por Mar y Tierra*, which later attracted attention among historians of graphics.[6] But how could he establish his priority without fully revealing the details of his method?

It was not uncommon in those days for scientists to claim priority for some discovery by publishing a description in code, thereby putting it "out there" without revealing the details. For example, in 1610, Galileo, with the aid of his first telescope, was making discoveries of the moon, Jupiter, and Saturn faster than he could write them up. On July 25, he dashed off a letter to Kepler (and others) with this coded description of his most recent discovery:[7]

> s m a i s m r m i l m e p o e t a l e u m i b u n e n u g t t a u i r a s

Kepler never solved this simple anagram cipher, but Galileo later revealed it as

> Altissimum planetam tergeminum observavi

meaning, "I have observed the highest of the planets [Saturn] three-formed." This referred to the peculiar shape of Saturn he observed, which would later be more accurately described by Christiaan Huygens as the rings of Saturn.

ImleV9 ap3Apa Ihrr5e tlSmelf9 5lesEortEr 5e eadnu9c Rtl9e9T omgupea Nſnnd cAlve-
Ma dſneagL p9rIir5 rEant tdTeo9Im nc5T9t noqCtuN veroQn nnmEef alarRl 9kle ral-
man Me4rn eqtlu u4xVeu ulriqDa ſuVne etſelId ſe5tſ couAu 9ſ9Vldu lir5te Tce40 vEe7oſ-
nE i5uameg Ebſe lodRa 9ebtſl Sa95u rVemai Aenprlt a9dL3do9 9nRt e3enqQe cun5ef
Etſot dEr 5emus Oeacdſae 5ucſoMe e9lrrI9 acnuoEd umr92 L5d9a5 eI9cnai dnneNt t4pA-
leai gPrmrO e5e VnſzbmF oaenſeS5 uſlOnt teoDe p9noll l9lo Enen trEge59 cut To 9u-
ned V9neq ItduLau Deum NamDe nEerEmſ9 9LmdVl eR99mEe e5nOu rdTd9 oOedu
l9oVa5 nqnp ntEaE eerlVrt lLrT9 5etoſ Y9ntl Sfrnae eG9a6 rſailau uulAnoTtp 9qVe rulr-
ſeT t9pOu erE9 leLſln Ecedo EſrNn eMeſu 3Nove Ar9ſ VmdtS qeVeueEd oVn9nuſu R9-
ſenPe utrTl 5eAten Afica qTe9u prSa a5trOl rle5ef hRſ95 eDluſ Iert5 eoVa ſ9qc lS u ela-
let eſ9Oſd qtuuef eſ9pero tmuaaru mumeuen yſtdm aeeuNr 9tlne eſnmſt pTdaſ 9n3t taMe
qnſutu euDalnſa depesE rſeedrm9 l9tVe5e ſrſaeu H9uia aſnſet tRefre ſe comſ9p ſtAle v9du
Qde95 3dLloe eu5ale uea4Rrfe ſ9l5na4 dAme 5nnr neoeſR nrtearo oe7uſOn uuoer9r pſtr
tEn9e rnresEa aoplna afrſa lSe9 Eecrſoae nTſſ4l teoolLt 9atlq elnr eeuſlCn elune e3frLo 97m-
neb 9tE9r teaena aduNue ſ4tſ9Ve ytm eepaNe ſnled9 lCln ladXedr ſS9eſ tſe5u uepulſ p9to-
dNo re9tnl etlpLe eaeſ rqeEurua aeE9alau qCnmu te5Snſ lom9t Ce5em gRoeenr dPl9ea
dNq9 9nTſeos nyMed 4ru9al ee9uoeE luuold ue uurdeD.

2.5 **Cipher:** Van Langren's cipher-text description of his solution to the longitude problem. This cipher has never been decoded. *Source:* M. F. van Langren, *La Verdadera Longitud por Mar y Tierra*, Antwerp, 1644.

Van Langren resorted to a similar stratagem to publish his idea yet keep it secret. In *La Verdadera* (p. 6) he says,

> Van Langren began the study of the Longitude for sea and land by means of the Moon in the year 1621. In 1625 he informed the Infanta Isabel about this method to calculate the Longitude, as well as a second method that Langren had discovered (that it is written later in dark letters), as it is known from a letter that infanta Isabel wrote to his Majesty in the same year of 1625.

The "dark letters" are a ciphered text that he includes, as shown in Figure 2.5. In 2009 this text was sent to a variety of amateur and professional code-breakers as a challenge. But as yet, no one has cracked the message, partly because it was written in old Spanish and uses both letters and digits, but also because the general form of the cipher (substitution, transposition, anagram) has to this day resisted attack.

The "Secret" of Longitude

Although van Langren's cipher remains undecoded, the "secret" he alluded to in his letters and *La Verdadera* can be inferred from his other work. The essential idea was to use recognizable *features* of the moon—rather than its mere position in the sky—to give a more accurate celestial clock. By timing the occurrence of sunrise or sunset on identifiable lunar peaks and craters, one would have a nearly continuous set of reference events with which local time could be accurately determined.

Two things were needed to make this idea practical for determining longitude at sea. First, it required an accurate lunar globe or set of maps that named the peaks, craters, and other lunar features so that they could be easily recognized. Second, it required a set of ephemeris tables, recording the onset in standard time of sunrise (lightening) and sunset (darkening) events on the days of the lunar cycle.

Assured of a salary and a position in the Spanish court, van Langren made plans for the preparation of a collection of lunar maps and diagrams, together with a "user guide" containing instructions for the calculation of longitude from observations of the lunar features that he intended to catalog. Because he would be the first to comprehensively map the lunar features, he proposed to have "the names of illustrious men applied to the luminous and resplendent mountains and islands of the lunar globe," a prospect that evidently pleased King Philip, because he and his initial patron, the Infanta Isabella, would appear many times in his nomenclature. Van Langren's first lunar map, titled *Plenilunii Lumina Austriaca Philippica* and dedicated to King Philip, showed 325 topographic names that he had assigned to lunar features.

Van Langren never completed the manual and tables describing exactly how his lunar map could be used. Moreover, although his scheme for longitude determination based on a detailed lunar map did offer the opportunity for greater precision than previous lunar methods, the relatively slow speed with which the lunar peaks became illuminated or vanished set hard limits on the precision that this method could achieve. To be sure, this was far better than observations once or twice a day, but nothing approaching the accuracy that was later achieved with a reliable marine chronometer. Nevertheless, he was the first to produce a comprehensive lunar map, and his own toponym for the crater *Langrenus* survives to this day, along with about half his other names.

Van Langren's Legacy

Today, Michael Florent van Langren is better known for his contributions to selenography—mapping lunar features—than he is for his contributions to solving the longitude problem in his time or to the development of data graphics.

Yet we hope you will agree that his invention of the one-dimensional dot plot is even today a stellar (or lunar?) example of clear visual presentation, from a time well before the ideas of uncertainty of observations or even of representation of empirical data values along an axis were contemplated.

Van Langren's personal life has also been enigmatic: no portrait of him exists and nothing was known of his family life or burial site. Recent research[8] has uncovered a wealth of new details: Michael married Jeanne de Quantere and they had four known children between 1626 and 1635; at age twenty-nine, he also had an illegitimate daughter with Jeanette van Deynze, whom he later acknowledged and legitimized in 1657. We now know from Wauters (1891, 1892) that Michael died in the first days of May 1675 and was buried on May 9, in the church of Notre-Dame de la Chapelle, Brussels. But by 1890, however, there was no trace of his burial there.

3

The Birth of Data

In one of the first published cookbooks, around 1860, Mrs. Isabella Beeton began her recipe for rabbit stew with the instruction: "First, catch a rabbit." So too an early, prescient recipe for data graphics might have begun, "First, get some data." The second step in the recipe might have been, "Now, make some sense of it!"

Slightly later (1891), Arthur Conan Doyle had Sherlock Homes proclaim in *Scandal in Bohemia*, "It is a capital mistake to theorize before one has data. Insensibly one begins to twist facts to suit theories, instead of theories to suit facts." These popular ideas set a theme for this chapter: the connections between observations, quantified as "data," and conclusions based on the evidence those observations provide, facilitated by graphs for discovery and communication.

The idea of deriving knowledge through observation and experience, as opposed to inner thought, starts, in Western tradition, with Aristotle's view that all knowledge comes through our sensory experience: our concepts of an *apple* or a *tree* are derived over time through numerous encounters with examples from which we learn the essential features. Aristotle made this idea concrete with the notion of the human mind as a blank slate (tabula rasa) on which experience records its marks.

But this idea did not really gain adherents until the rise of British empiricism (with John Locke, George Berkeley, and David Hume) and the Age of Reason in the seventeenth and eighteenth centuries. In part, the prior lack of empirical data accounts for the gap in innovation of graphical methods between van Langren in the early seventeenth century and the explosion of graphical methods in 1780–1840 as we saw in Figure 2.3.

The systematic and widespread collection of data developed steadily over this time in response to important issues in astronomy (the "shape" of the

Earth, orbits of planets), political economy (new markets, balance of trade), and social factors (literacy, crime). These and other areas provided the essential ingredients for Mrs. Beeton's recipe for graphs: just as gastronomy or hunger, together with availability of rabbits, may have driven the recipe for rabbit stew, so too did important scientific questions propel the collection of empirical data in order to refine concepts or test competing views against each other.

The general principles of starting with a well-defined question, engaging in careful observation, and then formulating hypotheses and assessing the strength of evidence for and against them became known as the *scientific method*. This chapter traces the role of data in the initial rise of graphical methods around the early 1800s, which can be called an Age of Data and which provided the "big data" of its time. There were certainly many cooks and sous-chefs in the data kitchen, whose contributions we describe. We focus attention on one important participant in this story: André-Michel Guerry [1802–1866], who used an "avalanche of data" and graphical methods to help invent modern social science.

To appreciate what was new in this period, when widespread data collection created a climate for the development of graphical methods, it is useful to make a distinction between the mere recording of numbers and what can be called "evidence," in the case where a collection of numbers is used to connect an idea, goal, or hypothesis to some conclusion, argument, or prediction.

Early Numerical Recordings

The recording of numbers that could be called "data" (under a loose definition) goes back to antiquity. For 7,000 years prior to construction of the Aswan High Dam, people lived and farmed along the Nile. One early and well-documented source are the records of the times and heights of the flooding of the Nile, which today is still celebrated in Egypt for two weeks starting August 15 as the holiday Wafaa El-Nil. When Herodotus began writing about Egypt and the Nile (circa 450 BCE), the Egyptians, who knew that their prosperity depended on the river's annual overflow, had been keeping records of the Nile's high water mark for more than three millennia. In 1951, Popper presented a time series of the Nile flood levels over thirteen centuries, from AD 622 to 1922, which was perhaps the longest time series ever recorded.

However, we should not consider this as evidence of any sort because there was no sense that what happened in past years could be considered as an *aggregate* collection of numbers you could do anything more generally useful with. If you were a farmer on the Nile, you probably knew the date and level of flooding for the last year or so. But this gave only a little help in deciding when to plant or whether you could afford to buy another ox five years later. The historical record, as detailed as it was, comprised just a collection of individual numbers that were seen through a close-up lens of the very recent past. Certainly no one thought to make a chart of the high water level over time or attempt to compare the average water level in the last decade to what might occur in the next.

Another old, and extremely detailed, source of numerical recording are the so-called ephemeris tables (from the Latin and Greek words meaning "diary" or "calendar"), giving the positions of astronomical objects (the moon, stars, and planets) in the sky at given geographic positions, at regular intervals of date and time. After his coronation in January 1252, King Alfonso X of Castile commissioned a new, more accurate and detailed set of tables, the Alfonsine Tables, giving data for calculating the position of the sun, moon, and planets relative to fixed stars (Figure 3.1). The first printed version did not appear until 1483.

Together with other sources, these were the data from which Nicolaus Copernicus crafted his heliocentric theory of the solar system in 1543. Quite soon after that, in 1551, Erasmus Reinhold published his Prutenic Tables, which he dedicated to his patron, Albert I, Duke of Prussia. Decades later (1627), Johannes Kepler published the Rudolphine Tables (dedicated to the Holy Roman Emperor Rudolph II), which would provide the source for Kepler's discovery of his laws of planetary motion. These and other sources were used by van Langren to compile the data plotted in his graph (Figure 2.1), and his aim for his "secret of Longitude" was to compile similar tables giving observations of sunrise and sunset for mountains and craters on the Moon.

Yet in spite of all this activity, we are still reluctant to call these collections of numbers evidence in the strong sense intended here, because, like the data on flooding of the Nile, few people thought to use them for purposes other than looking up a historical record or making a local calculation from isolated numbers. The individual observations of the times of rising of the Moon

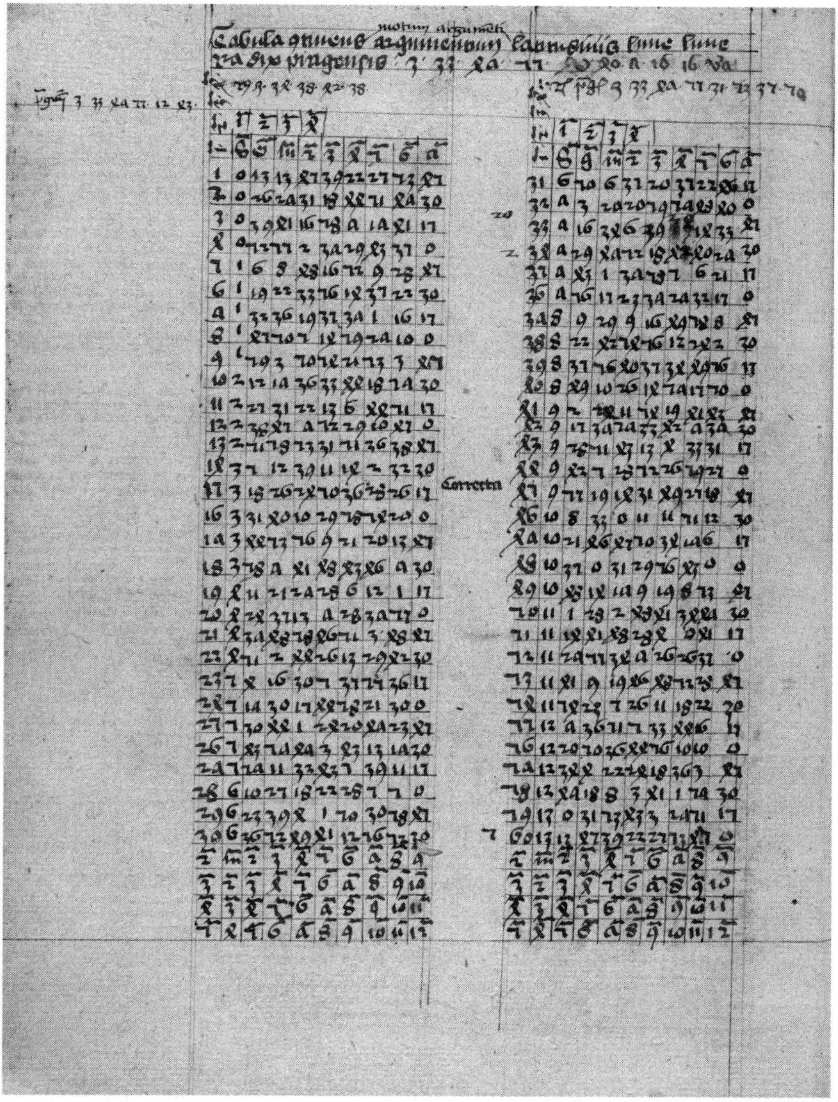

3.1 **Alfonsine Tables:** A page from the Alfonsine Tables giving times of observations of celestial events. *Source:* Kislak Center for Special Collections, Rare Books and Manuscripts University of Pennsylvania, LJS 174.

in Toledo or the elevation of the North Star or Betelgeuse on different dates recorded by Tycho Brahe and other astronomers were all certainly useful for navigation and for constructing theories of planetary motion, but they were not yet completely evidentiary data in the modern sense.

In 1601, Kepler acquired the meticulous catalogs of the celestials positions of planets and stars recorded by Tycho Brahe in his observatory on the Danish island of Hven. Brahe's observations were so precise that Kepler was able to calculate the orbit of the planet Mars accurately enough to be able to distinguish between a circle, a parabola, and an ellipse; by 1609, he was able to state the first of his three laws of motion: "The orbit of every planet is an ellipse with the sun at a focus." These observations were very nearly data in the modern sense.

Political Arithmetic

The first real inkling and widespread realization that data could be put to some larger and more general use occurred in 1662 when John Graunt [1620–1674], a London haberdasher by trade, published *Natural and Political Observations Made upon the Bills of Mortality*. Here, he laid out the first statistical estimates of the population of London, based on official birth and death records, and presented life tables giving survival numbers to each age. Together with his friend William Petty, he developed the idea that population numbers (later called demography) could be useful for a variety of purposes of the state, including taxation and how to raise an army, as well as for economic purposes, such as how to value annuities on a life or price insurance policies. The *Bills of Mortality* also gave rough categories of cause of death, so Graunt could show that deaths from chronic conditions outnumbered those from plague and other epidemics that caused great fear at this time.

We consider this event the birth of "data" as empirical evidence in our modern sense: numerical facts began to be viewed not as individual elements (that could be looked up, compared, and even calculated with) but rather as constituent members of an *ensemble*, an aggregate collection of similar numbers organized to support or refute claims for some larger purposes. This event was also the beginning of the study of "social numbers," the numerical characteristics of individuals in human society regarded in the aggregate.

But it was William Petty [1623–1687] who fully appreciated the use to which such data could be put. From humble beginnings in a family of London clothiers, Petty become a professor of anatomy at Oxford University in 1650, and a chief physician to Oliver Cromwell's army in Ireland in 1652, before being appointed a founding member of the Royal Society in 1660 and knighted a year later by Charles II.

If Graunt can be said to have invented the idea of "data" as evidence, Petty can lay claim to the invention of data analysis in various writings between 1685 and 1690 on a subject he called "Political arithmetik." It was based on the simple idea, well-known to shopkeepers, of standardizing raw numbers by the relevant totals and the "rule of three," *a* : *b as c* : *?*, to make proportional comparisons. From this one could calculate any one number from the others. For example, one could calculate how many hats (*?*) might be sold this year, given the numbers of suits sold this year (*a*) and last (*b*), and the number of hats (*c*) sold last year; one could also continue this calculation over time to see how sales progressed from year to year.

Thus, political arithmeticians were first able to establish a rational basis for comparisons over time, age, geographic region, and other categories. In *Political Arithmetic* (c. 1676, published 1690) Petty applied these ideas to the study of manufacturing, agriculture, and other aspects of the economy and public life, wherever he could get his hands on some numbers. These developments are often considered the birth of what we now call statistics, although that term (*statistik*, meaning "numbers of the state") was not introduced until 1749, by Gottfried Achenwall. It is in these developments that we place the birth of the idea of "data" in our modern sense.

The Human Sex Ratio

The *Bills of Mortality* led to some something larger and more general in the history of numbers as evidence for a proposition. Graunt's data were based on parish records of christenings and deaths, recorded nearly weekly and with at least a modicum of uniformity. In 1710, John Arbuthnot [1667–1735], a Scottish minister and physician to Queen Anne, calculated the ratio of male to female births from these records for the period 1629–1710. He was amazed to observe that the ratio was *always* greater than 1 (see Figure 3.2), even if only slightly. If male and female births were equally likely, this result would

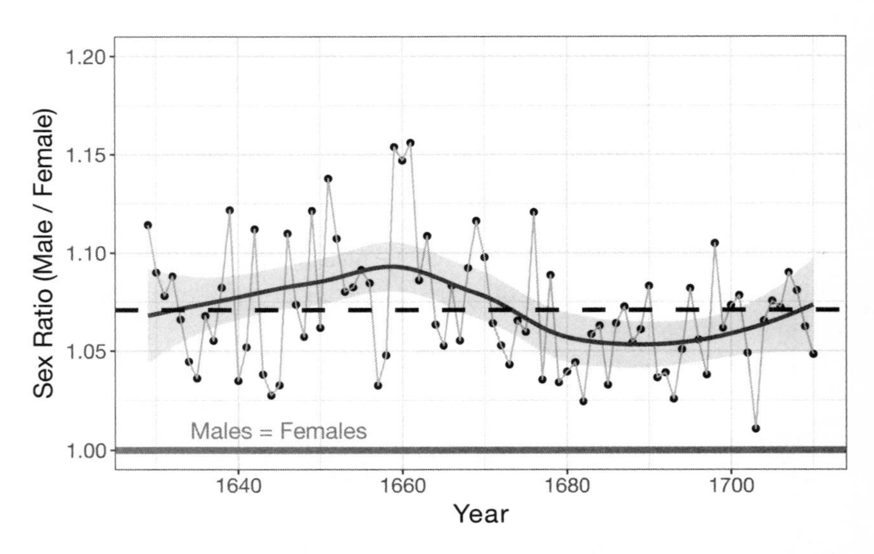

3.2 **Sex ratio:** Arbuthnot's data on the ratio of male to female births. The average ratio, 1.07, is shown by the upper dashed line, compared to the solid line at 1.0. The curved line shows a smoothed (loess) curve through the points, with a shaded confidence interval. *Source:* © The Authors.

be like tossing 82 coins and obtaining all heads, which has a probability of $(\frac{1}{2})^{82} = 2 \times 10^{-25}$, a very small number.

Arbuthnot used this apparently lawful regularity to argue that Divine Providence, not chance, governs the human sex ratio.[1] His argument was perhaps the first application of probability to social statistics; this can be considered the first formal significance test of a statistical hypothesis.

His conclusion was wrong because the larger number of males can now be at least partially attributed to higher female fetal mortality before birth, not to mention factors that might intervene between birth and christening.[2] Yet, a set of numbers had finally been converted to evidence for an argument, and Arbuthnot provided an initial idea for how to measure the *strength* of evidence.

Graunt's data later contributed to the development of *inverse probability* (now called Bayesian theory) by Pierre-Simon Laplace [1749–1827], by which the a posteriori probability of a hypothesis could be calculated from a series of observations, thus giving the most probable causes of known events. Searching for a sufficiently large database, Laplace turned to the sex-ratio problem using Graunt's data for London and similar data collected for Paris. In several papers published between 1774 and 1786, he established a rationale for the conclusion that the probability of a male birth is almost certainly greater than 1/2, and he stated the degree of his conviction as a bet that boys would outnumber girls for the next 179 years in Paris and the next 8,605 years in London. Such was the power of numbers and theory.

The evidence was so strong for a small, but systematic, bias in favor of male births that neither Arbuthnot nor Laplace would have seen any need to make a graph like Figure 3.2. Yet the overwhelming visual evidence of such a graph might have made proof based on numerical calculations unnecessary.

The next step in the story of data was to connect numbers to visual displays as an aid to reasoning about evidence for or against a theory or hypothesis. This was soon to come, but it required more data on important social issues.

An Avalanche of Numbers

By the mid-1700s, the importance of measuring and analyzing population distributions was recognized and the idea that ethical and state policies could encourage wealth through population growth was established, most notably in 1741 by Johann Peter Süssmilch [1707–1771]. Süssmilch advocated the expansion of government collection of population statistics and pointed to a nearly constant sex ratio favoring males as one illustration of the regularity to be found in official data. He is regarded as one of the founders of demography and a pioneer in the history of population statistics. Data on the social character of human populations was still lacking, however.

This would soon change. The first impetus for the *widespread* systematic collection of social data occurred in France in the period following the French Revolution to the fall of Napoleon in 1814 and onward to the Bourbon Restoration, ending in July 1830. This period saw widespread inflation, unemployment, and upheaval. Paris witnessed explosive growth and food and housing shortages, accompanied by what was perceived in the Parisian press as a vastly increased rise in crime, which was attributed to a new, dangerous class of petty criminals (Chevalier, 1958).

Crime was not the only social issue. The Poor Laws in England subjected the poor to almshouses and debtors to workhouses or debtors prisons; the condition of being indigent was termed *pauperism*, the peculiar English term likening destitution to a disease or chronic condition. Suicide became an Anglo-French issue after George Burrows noted that in 1813 there were 141 suicides in Paris but only 35 in London, as well as 243 drownings in the Seine compared with 101 in the Thames, most of which he considered to have been "voluntary death."[3]

There was much debate about the treatment of prisoners in the popular press, journals, and learned societies. Then, as now, there were two basic schools of thought on criminal justice policy: a liberal *philanthrope* position advocated increased education, religious instruction, improved diet (bread and *soup!*) and better prison conditions as the means to reduce crime and recidivism. On the other hand, hard-line conservatives feared attempts at prison reform, doubted the efficacy of campaigns for public education, and viewed suggestions to abandon the harsh punishment of convicts with grave suspicion, if not alarm. But the evidence marshaled to support such armchair recommendations was fragmentary, restricted, and often idiosyncratic, as when a sensational murder in London or Paris occasioned new calls for attention to the issue.

The dearth of satisfactory data on these issues began to change in Paris with annual publications of the *Recherches statistiques sur la ville de Paris et le départment de la Seine* under the direction of the mathematician Jean Baptiste Joseph Fourier [1768–1830], starting in 1821, toward the end of his life. These volumes detailed births, marriages, and deaths, but they also provided extensive tabulations and breakdowns of inmates of Parisian insane asylums and motives and causes of suicides.

With Fourier's tabulations as a model, in 1825 the French Ministry of Justice began the first centralized, national system of crime reporting, with data collected quarterly from every department in France. It required recording the details of every criminal charge laid before the French courts: age, sex, and occupation of the accused; the nature of the charge; and the outcome in court. Annual statistical publications of these data, known as the *Compte général de l'administration de la justice criminelle en France*, began in 1827 under the initiative of Jacques Guerry de Champneuf, the director of *affaires criminelles* in the Ministry of Justice.

During the same period, a wealth of other data on moral and other social variables became available from other sources: data on age distributions and immigrants in Paris began with the 1817 census; Alexandre Parent-Duchâtelet (1836) provided comprehensive data on prostitutes in Paris by year and place of birth; the Ministry of War began to record data on conscripts' ability to read and write; information on wealth (indicated by taxes), industry (indicated by patents filed) and even wagers on royal lotteries became available for the departments of France in various bulletins of the Ministry of Finance in 1820–1830.

Thus, the first step in Mrs. Beeton's recipe was achieved by what Ian Hacking[4] rightly called an "avalanche of numbers." It remained for someone to try to make sense of competing claims about the causes of crime and relationships among moral variables from this comprehensive data and detailed analysis.

In some initial steps, data on crime from the Compte général of 1825–1827 was combined with data from the census to give standardized measures of population per crime (number of inhabitants for one condemned person) for the eighty-one departments of France. Data on school instruction, as another measure of literacy, were based on the number of male children in primary schools in twenty-six educational districts in France, also in the form of inhabitants per student. A credible basis for comparing social numbers over time and geographic space had been established. But it remained to visualize such data and reason to credible conclusions.

Mapping Social Data

The idea of displaying quantitative data on a map—now called thematic cartography—has a long history.[5] An early example, generally considered the first thematic map of economic data,[6] is the *Neue Carte Von Europa* (1782)[7] by the German economist August Friedrich Wilhelm Crome [1753–1833]. Crome used a variety of iconic symbols to show fifty-six commodities and products: gold, silver, cows, fish, tobacco, and so on. Yet, while this was an important landmark in map-based data display, it didn't "speak to the eyes." In other words, one could look up where wine was produced yet not see any pattern nor connect the dots to understand any connection with terrain or climate.

In an important step toward visualizing and understanding map-based data, Baron Charles Dupin [1784–1873] had the idea to portray the levels of instruction (proportional number of male children in primary schools by department) as degrees of shading on a map of France. His 1826 map used graduated shadings from black to white with darker shades representing increasing degrees of illiteracy or ignorance, making this the first known instance of what today is called a *choropleth* map.

It immediately became clear that a diagonal line from Brittany to Geneva separated the less educated south of France from the better educated north, a distinction discussed for many years as *France obscure* versus *France éclairée*. This invention—the first modern statistical map—was the starting point of a true graphical revolution that would soon extend to a more general social cartography with the comparative analysis of social issues.

With so much data, the arm-chair philosophizing on the relations of crime with other variables could now be addressed directly. In 1829, André-Michel Guerry, a young lawyer working in the Ministry of Justice, teamed up with the Venetian geographer Antonio Balbi to produce the first statistical map of crime data (Figure 3.3).

The concept of correlation would not be discovered for another sixty years (by Francis Galton in 1886), so Guerry and Balbi did the next best thing: in another first in statistical graphics, they presented maps of instruction, crimes against persons, and crimes against property in a single large sheet. Tufte[8] has dubbed this idea "small multiples" because it allows *direct* visual comparisons of related sets of data, which are usually subsets presented adjacently in the same display. This serves the purpose of "visually enforcing comparisons of changes, of the differences among objects, of the scope of alternatives."[9]

The maps in Figure 3.3 were surprising to both the liberal and conservative camps because they went far beyond the ken of rationalist arm-chair philosophy and contradicted parts of each view. The comparative maps showed that (a) personal crimes and property crimes seemed inversely related overall, but both tended to be high in more urban areas; (b) the clear demarcation between the north and south of France in instruction was even more pronounced than in Dupin's map; and (c) although the north of France had the highest levels of education, property crime was also high there. At the very least, this work signaled that "crime" could not be considered a unitary construct and testified to the importance of using detailed data, sensibly presented, to inform the debate on the relations of crime and education.

Other applications of this graphical method in the social realm soon appeared in France, the Netherlands, England, and elsewhere in Europe, as the general study of "moral statistics" took shape and assumed wider scope.

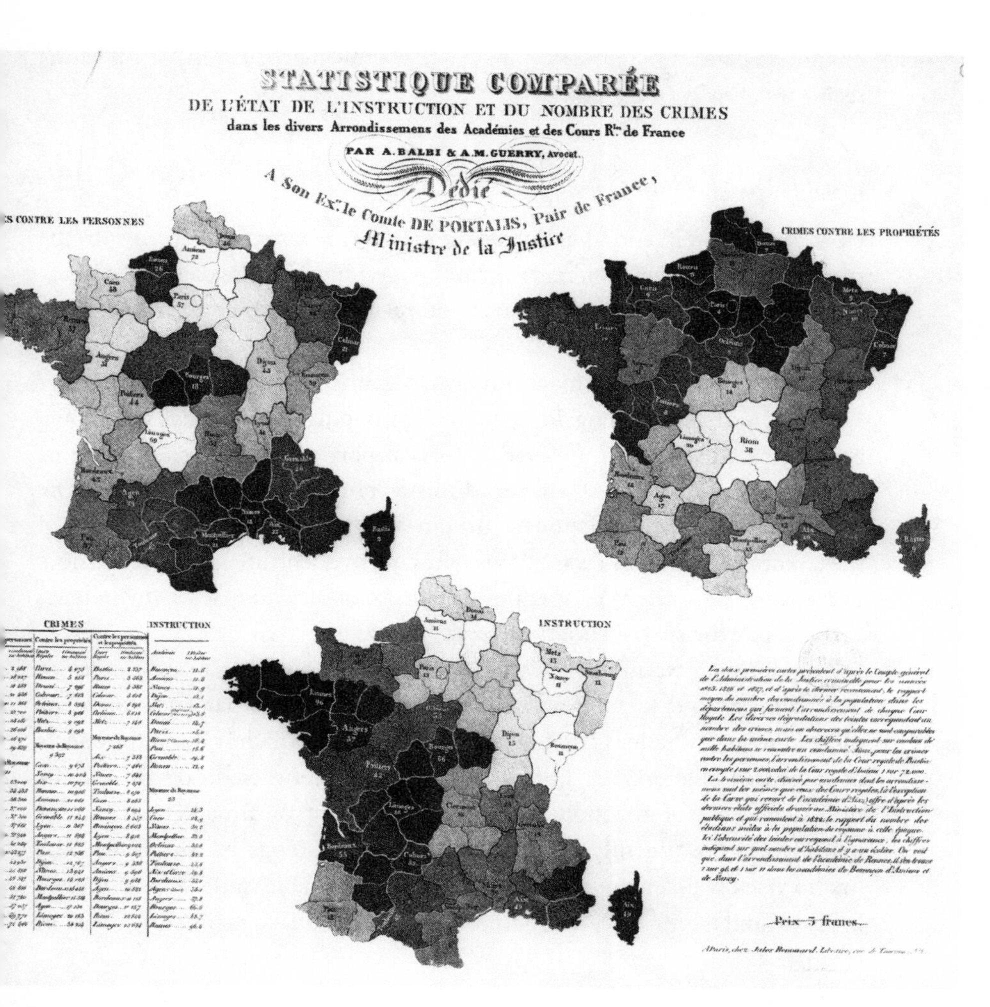

3.3 **First comparative shaded maps:** Guerry and Balbi's 1829 *Statistique comparée de l'état de l'instruction et du nombre des crimes.* Top left: crimes against persons; top right: crimes against property; bottom: instruction. In each map, the departments are shaded so that darker is worse (more crime or less education). *Source:* Reproduction courtesy of the Bibliothéque National Français.

These shaded maps dealt with education, criminality, begging, prostitution, poverty ("pauperism"), suicide, and other social topics. For instance, Alexandre Parent-Duchâtelet (1836) showed the distribution of birthplace of 12,200 Parisian prostitutes in Paris by department of France and by arrondissement of Paris. In the Netherlands, Hartog Somerhausen (1829) applied Dupin's method to education in the Pays-Bas. The graphic ingredient in Mrs. Beeton's recipe was finally in the pot.

Graphic Details Matter

However, just as in cooking, the details matter: the wrong spice can ruin the stew. In graphing data, different methods or graphical features can make it easier or harder to perceive and understand relationships or comparisons from the same data.

Shortly after Dupin, Armand Frère de Montizon (1830) invented another cartographic method using dot symbols. The population of France was shown by departments, using a number of dots proportional to the number of inhabitants, 1 dot to 10,000. He called this a "philosophical map" because he wished to relate the population to "the physical, intellectual and moral state of the country." Yet his map was not visually effective. The image appeared more uniform because the very small dots and the observation scale (by department) made the spatial variations less evident.

The dot map would later come to attention in showing disease cases or deaths linked to epidemics on a large scale, often in an urban context.[10] The most famous example is Dr. John Snow's 1855 map, which showed the effects of cholera in a London neighborhood and the link between the deaths and the probable source of infection, the public water pump on Broad Street. Snow's map figures prominently in the history of epidemiology because it was the first to visually link an outbreak of disease to a probable cause—the cluster of cases around the Broad Street pump. We return to this topic in more detail in Chapter 4.

Somewhat later, Adolphe Quetelet (1836) presented two comparative maps of crimes against persons and against property in France in a single sheet. Quetelet introduced yet another method to show geographic variation of moral variables: using continuous shading across internal boundaries, rather than uniform shading within each department. His resulting maps, shown in Figure 3.4, may have been visually appealing, but we don't find that they lead as easily to comparisons and conclusions as Guerry's use of choropleth maps.

These graphic variations of thematic maps have continued to be debated and refined to the present day. Their importance in the historical context is that a link between data, an image (map), and a scientific question had been

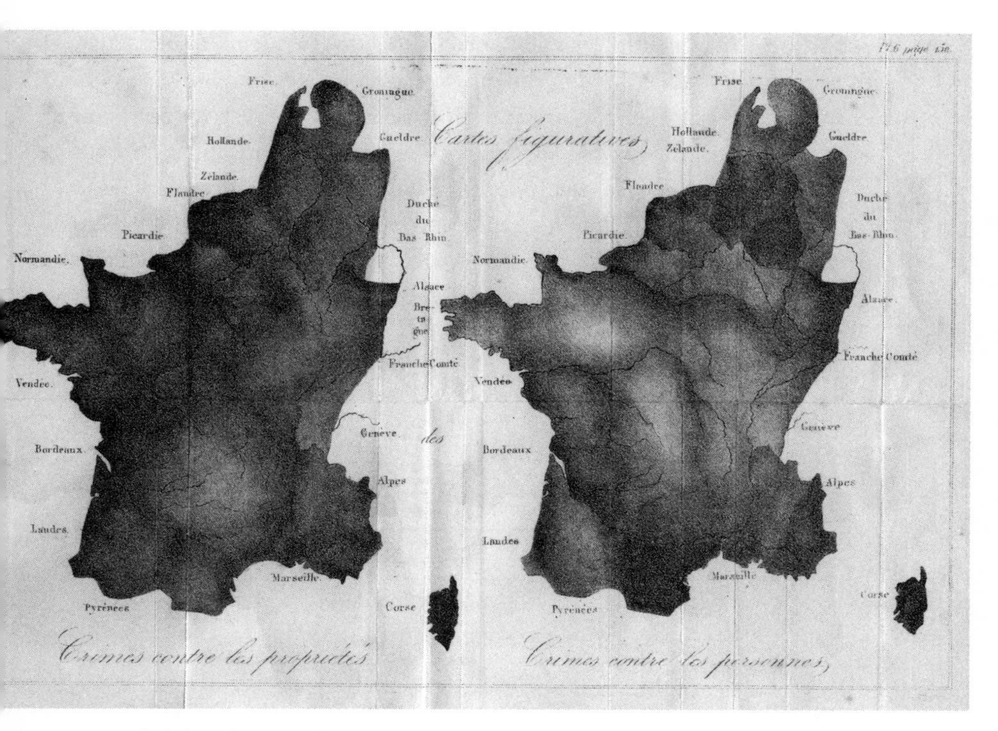

3.4 **Quetelet's (1836) maps of crime:** Crimes against property (left), crimes against persons (right). *Source:* Reproduction courtesy of Princeton University Library, Historic Maps Collection.

established. One could then attempt to reason about pressing social questions by means of graphic displays—data had become evidence, shown in a graph.

Stability and Variation

The next breakthrough in the Age of Data took place only three years after Balbi and Guerry's initial comparative map. On July 2, 1832, Guerry presented a slim manuscript to the Académie Française des Sciences titled *Essai sur la statistique morale de la France* (henceforth *Essai*), in which he assembled the available data from 1825 to 1830 on crimes, suicides, literacy, and other moral variables and used proportional tables and maps to analyze these social issues.

His method was simple: he tabulated the relative frequencies (percentages) of crime and suicide, broken down by geographic region, age, sex, type of crime or method of suicide, and month or season of the year, for each year of available data. Some sample results are shown in Tables 3.1 and 3.2.

The visualization of his data was immediately compelling, even when presented in table form: the rates of crime (and suicide) remained remarkably stable over time, no matter how broken down, but they varied systematically by region, sex of accused, type of crime, and even season of the year. In any given French department or region, almost the same proportions of inhabitants stole, committed indecent assault, gave birth out of wedlock, and so forth.

Table 3.1. Tables from Guerry (1833, p. 11) showing percentages of crimes by several characteristics. The percentages remain remarkably constant over years.

Year	**1826**	**1827**	**1828**	**1829**	**1830**	**Avg**
Sex			**All accused (%)**			
Male	79	79	78	77	78	78
Female	21	21	22	23	22	22
Age			**Accused of theft (%)**			
16–25	37	35	38	37	37	37
25–35	31	32	30	31	32	31
Crime			**Committed in summer (%)**			
Indecent assault	.	36	36	35	38	36
Assault & battery	.	28	27	27	27	28

Table 3.2. Table from Guerry (1833, p. 10) showing percentages of crimes against property by year and region of France.

Year	**1825**	**1826**	**1827**	**1828**	**1829**	**1830**	**Avg**
Region			**Crimes against property (%)**				
North	41	42	42	43	44	44	42
West	17	19	19	17	17	17	18
East	18	16	17	16	14	15	16
Central	12	12	11	12	13	13	13
South	12	11	11	12	12	11	12
Total	100	100	100	100	100	100	100

This combination of stability (over factors that don't matter, like year) and variation (over variables that should, like type of crime) gave rise to the idea of the law-like behavior of social facts. These results led Guerry to ask whether crime and other moral variables are simply indicants of individual behavior or whether human actions in the social world are governed by social laws, just as inanimate objects are governed by laws of the physical world. This was a revolutionary idea. He argued: "Each year sees the same number of crimes of the same degree reproduced in the same regions. . . . We are forced to recognize that the facts of the moral order are subject, like those of the physical order, to invariable laws" (Guerry, 1833, p. 10, 14).

In passing, we note that Adolphe Quetelet [1796–1874] published a paper in 1831 on the development of the propensity to commit a crime in which he described similar analyses, relating crime to various social factors. This occasioned a priority dispute[11] for the discovery of lawfulness in social data. Here, we emphasize the role of Guerry, who was far less well-known, but who developed ideas of data analysis, graphic displays, and visual explanations to new levels.

Seeking Explanations, Causes, and Relationships

Guerry's 1833 *Essai* contained numerous tables giving breakdowns of crimes against persons and property by characteristics of the accused, frequencies of various subtypes of crime in rank order for both men and women (the most common personal crime for men was assault and battery, and for women, it was infanticide, often arising from unwanted pregnancy), and frequencies of crimes by age groups. To go beyond simple description, Guerry classified the crimes of poisoning, manslaughter, murder, and arson according to the apparent motive indicated in court records. For instance, in the case of poisoning, the motive was most frequently adultery; for murder, it was hatred or vengeance. This was a crucial step in understanding and explaining criminal behavior. It pointed to the need to study the relations among moral variables in new ways.

This quest to examine motives and causes is most apparent and impressive in Guerry's analysis of suicide, a topic of considerable debate in both the medical community (which considered it in relation to madness and other maladies) and the legal community (which considered whether it should be a crime, or at least within the purview of the justice ministry). "What would be useful to know would be the frequency and importance of each of these causes relative to all the others. Beyond this, it would be necessary to determine whether their influence . . . varies by age, sex, education, wealth, or social position."

To this end, Guerry carried out perhaps the first content analysis in social science by classifying the suicide notes in Paris according to motives or sentiments expressed for taking one's life.[12] But it irked him that, although crime was now routinely recorded in rich detail throughout France, outside Paris only the gross totals of suicide were recorded, without further details. In 1836 he began to create a system for the Ministry of Justice to mandate that local police record all details of suicides (demographic: age, sex, marital status . . .; social class or profession, literacy, moral character, . . .). Over the rest of his lifetime he personally examined over 85,000 suicide records collected between 1836 and 1860[13] and tried to tabulate them according to various potential causes and the imputed motive for suicide.

The *Essai* also contained a collection of bar graphs, highlighting certain comparisons: crimes against persons occurred most often in summer months, while those against property were most frequent in the winter; suicides by young males were most often carried out with a pistol, while older males preferred hanging (Figure 3.5). These simple graphs showed that crime and suicide were more nuanced than had been previously thought. Understanding them would require taking several potential "causes" into account together.

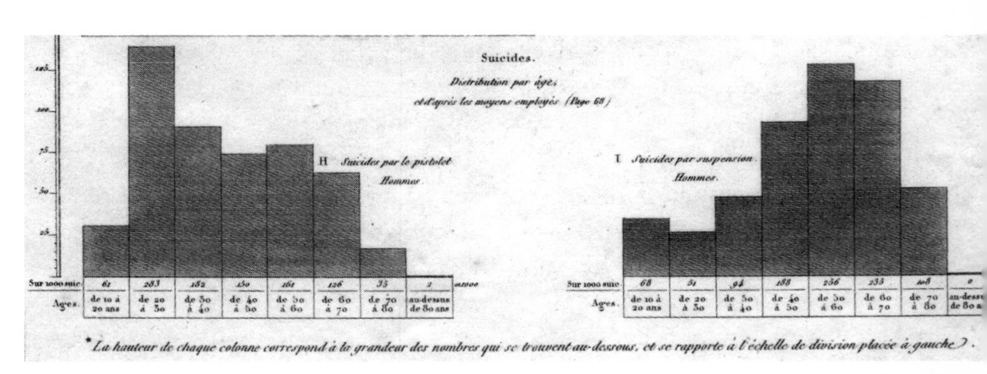

3.5 **Suicide method and age:** Histograms of the age distributions of suicide by pistol vs. by hanging, for males. *Source:* André-Michel Guerry, *Essai sur la statistique morale de la France*. Paris: Crochard, 1833, Plate VII.

INFLUENCE DE L'AGE.

DISTRIBUTION DES CRIMES AUX DIFFÉRENS AGES, PAR PÉRIODE DE DIX ANNÉES. IV.

A. CRIMES CONTRE LES PERSONNES.

…SSOUS DE 21 ANS.		DE 21 A 30.		DE 30 A 40.		DE 40 A 50.		DE 50 A 60.		DE 60 A 70.		AU-DESSUS DE 70 ANS.	
…RE DES CRIMES.	Sur 1,000	NATURE DES CRIMES	Sur 1,000	NATURE DES CRIMES.	Sur 1,000	NATURE DES CRIMES.	Sur 1,000	NATURE DES CRIMES.	Sur 1,000	NATURE DES CRIMES.	Sur 1,000	NATURE DES CRIMES.	Sur 1,000
…ures et coups.	184	Blessures et coups.	218	Blessures et coups.	179	Assassinat.	194	Meurtre	185	Meurtre.	173	Viol sur des enfans.	318
…sur des adultes.	169	Meurtre.	157	Assassinat.	154	Blessures et coups.	184	Assassinat.	182	Viol sur des enfans.	166	Blessures et coups.	137
…rtre.	147	Assassinat.	120	Meurtre.	152	Meurtre.	133	Blessures et coups.	175	Assassinat.	159	Meurtre	125
…sur des enfans.	123	Rébellion.	111	Rébellion.	110	Rébellion.	100	Rébellion	98	Blessures et coups.	138	Assassinat	102
…ssinat.	101	Viol sur des adultes.	105	Viol sur des adultes.	73	Viol sur des enfans.	94	Viol sur des enfans.	88	Faux témoignage. .	99	Faux témoignage. .	102
…llion.	78	Infanticide.	83	Infanticide.	63	Faux témoignage. .	69	Faux témoignage. .	76	Rébellion	78	Rébellion	91
…ticide.	48	Viol sur des enfans.	58	Viol sur des enfans.	59	Viol sur des adultes.	61	Viol sur des adultes.	32	Infanticide	42	Empoisonnement. .	23
. env. ascend.	47	Bless. env. acend.	50	Bless. env. ascend.	59	Bless. env. ascend.	44	Infanticide.	24	Empoisonnement, .	35	Infanticide.	23
…ciat. de malfait.	32	Faux témoignage. .	33	Faux témoignage. .	49	Infanticide.	41	Empoisonnement. .	20	Parricide.	21	Viol sur des adultes.	23
… témoignage. .	29	Empoisonnement . .	16	Empoisonnement. .	25	Empoisonnement . .	29	Bless. env. ascend.	19	Avortement	18	Associat. de malfait.	11
…oisonnement .	11	Crim. env. des enf.	10	Crim. env. des enf.	16	Associat. de malfait.	19	Avortement	15	Viol sur des adultes.	14	Voies de fait, etc. .	11
… de fait, etc. .	8	Associat. de malfait.	10	Associat. de malfait.	12	Bigamie.	15	Bigamie.	15	Crim. env. des enf.	11	Bless. env. ascend.	»
…l. av. violence.	6	Parricide.	8	Parricide.	9	Mendicité	8	Mendicité.	13	Bless. env. ascend.	7	Parricide.	»
. env. des enf.	5	Voies de fait, etc. .	6	Mend. av. violence.	8	Crim. env. des enf.	7	Parricide.	10	Associat. de malfait.	7	Crim. env. les enf.	»
…cide.	5	Avortement	5	Bigamie.	8	Avortement	7	Associat. de malfait.	10	Bigamie	7	Avortement	»
…tement.	3	Mend. av. violence.	2	Avortement	6	Parricide.	4	Crim. env. des enf.	6	Voies de fait, etc. .	7	Bigamie	»
…mie.		Bigamie.	1	Voies de fait, etc. .	6	Voies de fait, etc. .	4	Voies de fait, etc. .	6	Mend. av. violence.	7	Mend. av. violence.	»
…s crimes. . . .	4	Autres crimes. . . .	7	Autres crimes. . . .	12	Autres crimes. . . .	12	Autres crimes. . . .	26	Autres crimes. . . .	11	Autres crimes. . . .	34
TOTAUX. . . .	1,000		1,000		1,000		1,000		1,000		1,000		1,000

3.6 **Ranked lists:** Ranking of crimes against persons in seven age groups. Connecting lines show some noteworthy trends. *Source:* André-Michel Guerry, *Essai sur la statistique morale de la France.* Paris: Crochard, 1833, Plate IV.

In another novel graphic approach to questions of relationships and possible causes, in the 1833 *Essai* Guerry tried to examine how the types of crimes committed varied with age of the accused. To do this, he prepared the ranked lists shown in Figure 3.6 for crimes against persons sorted from high to low for each age group; a similar display showed rankings for crimes against property by age group. To make the trends more amenable to visual inspection, he added lines to connect selected crimes horizontally to show the trend. This gives a semigraphic display that combines a table (showing actual numbers) with the first known instance of a parallel coordinate plot. In the original, the trace lines are hand colored in different light hues to make them visually distinct. This is likely the first use of this combination of ranked lists showing data values connected by lines to show trends.

Using this display, Guerry discusses a variety of trends, such as the decrease in indecent assault on adults *(viol sur des adultes)* with age, the rise of indecent assault on children to the top for those over 70, and the increase in parricide with age (surprisingly reaching a maximum for "children" aged 60–70).

Additionally, to be able to discuss geographical differences and relate these moral variables to one another, he prepared six thematic maps of France (Figure 3.7), adding illegitimate births *(infants naturelles)*, donations

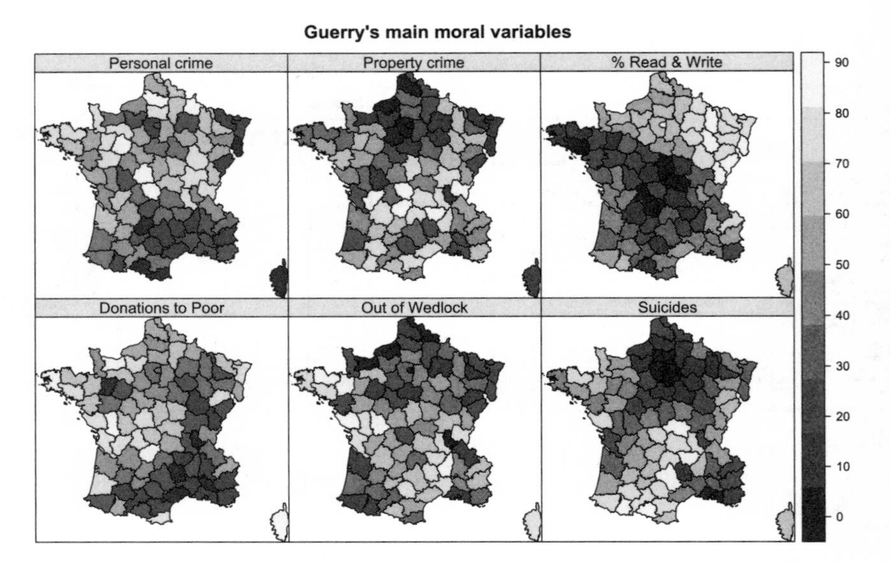

3.7 **Reproduction of Guerry's six maps:** As in Guerry's originals, darker shading signifies worse on each moral variable. Numbers give the rank order of departments on each variable. *Data source:* André-Michel Guerry, *Statistique morale de l'Angleterre comparée avec la statistique morale de la France, d'après les comptes de l'administration de la justice criminelle en Angleterre et en France, etc.* Paris: J.-B. Baillière et fils, 1864.

to the poor (number of gifts and bequests), and suicide to maps of personal crime property crime, and education presented earlier, but based on more complete data and better indicators.

Guerry conceived of another data and graphic innovation here: standardization of variables on topics with scales and their visual representation. First, he transformed his data to consistently ordered numbers in a form to express that "more is better." Therefore, he used percentage able to read and write to represent education, but an inverse scale of population per crime rather than crime per population. Second, in preparing the maps, these variables were converted to ranks and the departments were shaded by rank, so that darker tints were applied to the departments that fared worse on a given measure (e.g., more crime, less education). This transformation has the advantage that it is possible to compare the maps on these disparate topics and see where the dark areas (which fare badly) or light areas (which fare well) on two or more variables coincide. For example, the departments in the north of France are generally unfavorable on crimes against property, illegitimate birth, and suicide.

Guerry's *Essai* was received with considerable enthusiasm in European statistical circles, particularly in France and England. In France it was awarded the prestigious Monynton Prize from the French Academy of Science in 1833. Guerry was also elected to the Académie des sciences morales et politiques and at some point was awarded the cross of Chevalier of the Legion of Honor (Diard, 1867). He was invited to display the maps in several expositions in Europe, and, in 1851, had two exhibitions in England—an honored public one in the Crystal Palace at the London Exposition and a second at the British Association for the Advancement of Science in Bath. Guerry had literally put the intellectual discussion of social statistics on the map.

Analytical Statistics

Guerry's most ambitious work, and the capstone of his career, did not appear for another thirty years, in 1864. The *Statistique morale de l'Angleterre comparée avec la statistique morale de la France* was published in a grand format (56 x 39 cm, about the size of a large coffee table) and contained an introduction of sixty pages and seventeen exquisite color plates.

The introduction sets out Guerry's view of the history of the application of statistics to the moral sciences. In it, he proposes to replace the term *moral statistics*, or simply *documentary statistics*, with *analytical statistics*. The former, which is presented almost invariably in tables, is concerned with the numerical exposition of facts. The latter presents the successive transformation of these facts, by calculation, by concentration, and their reduction to a small number of general abstract results. One can see here a thorough explanation of the graphic method applied to moral and social data in the context of his time.

One cannot fail to be impressed by the sheer volume of data summarized in Guerry's graphic maps for different aspects of crime and other moral variables in France and England. These include over 226,000 cases of personal crime in two countries over twenty-five years and over 85,000 suicide records, classified by motive. Guerry estimated that if all his numbers were written down in a line, they would stretch over 1,170 meters! Hacking (1990, p. 80) credits this observation as the source of his phrase "an avalanche of numbers."

Re-Visions: Consulting for Guerry

In closing this chapter, it is worth asking how later developments might have made Guerry's task easier. Although he was mainly interested in understanding how crime and other moral outcomes were related to possible explanations, no statistical or graphical tools were available for such questions. As we will see in Chapter 6, nearly simultaneously with Guerry's *Essai*,

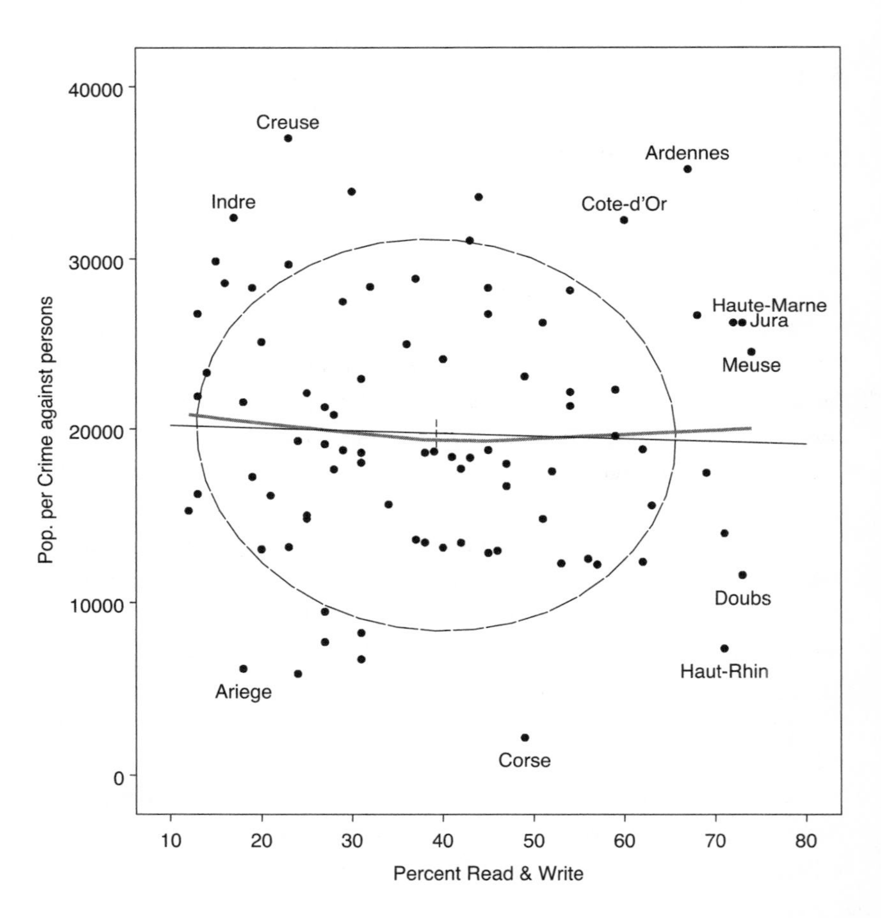

3.8 **Enhanced plot:** Scatterplot of crimes against persons versus literacy from Guerry's data, one point for each department. The black line shows the linear regression relation; the gray slightly curved line shows a nonparametric smoothing. Points outside a 90 percent data ellipse are identified by department. *Source:* © The Authors.

the British astronomer John F. W. Herschel invented the idea of plotting one variable against another, in what is now called a scatterplot. Sixty years later, Francis Galton formulated the idea of linear regression for quantifying the relationship between variables. This would have completed the infrastructure Guerry needed for his plan for finding associations among moral variables.

So, what could we do today if the young Guerry arrived at our door as a consulting client and asked for assistance in understanding his data on moral statistics of France? Our first suggestion would certainly be: make a scatterplot; enhance it to facilitate interpretation and presentation of the results. Figure 3.8 shows one such plot for the relationship of crimes against persons to literacy. To the basic plot of points, we added: (a) a linear regression line (black), (b) a smoothed curve (loess), and (c) a data ellipse containing the 68 percent of the points closest to the means of both variables. Points outside the analogous 90 percent data ellipse are labeled by department.

Thus, Guerry and his readers would have been able to see *directly* that overall there is no linear relation between crimes against persons and literacy, nor is there any hint of a nonlinear one. The departments that are labeled would have served to highlight the discussion along the lines that Guerry chose (e.g., the Ariége is near the bottom on crime and also on literacy, and Indre stands about the same on literacy, but is near the top on personal crime).

4

Vital Statistics

William Farr, John Snow, and Cholera

In the previous chapter we explained how concerns in France about crime led to the systematic collection of social data. This combination of important social issues and available data led Guerry to new developments involving data display in graphs, maps, and tables.

A short time later, an analogous effort began in the United Kingdom, in the context of social welfare, poverty, public health, and sanitation. These efforts produced two new heroes of data visualization, William Farr and John Snow, who were influential in the attempt to understand the causes of several epidemics of cholera and how the disease could be mitigated.

In the United Kingdom the Age of Data can be said to have begun with the creation of the General Register Office (GRO) by an Act of Parliament in 1836.[1] The initial intent was simply to track births and deaths in England and Wales as the means of ensuring the lawful transfer of property rights between generations of the landed gentry.

But the 1836 act did much more. It required that every single child of an English parent, even those born at sea, have the particulars reported to a local registrar on standard forms within fifteen days. It also required that every marriage and death be reported and that no dead body could be buried without a certificate of registration, and it imposed substantial fines (10–50£) for failure in this reporting duty. The effect was to create a complete data base of the entire population of England, which is still maintained by the GRO today.

The following year, William Farr [1807–1883], a 30-year-old physician, was hired, initially to handle the *vital registration* of live births, deaths, marriages, and divorces for the upcoming Census of 1841. After he wrote a chapter[2] on "Vital statistics; or, the statistics of health, sickness, diseases, and death," he was given a new post as the "compiler of scientific abstracts," becoming the first official statistician of the UK.

Like Guerry at the Ministry of Justice in France, Farr had access to, and had to make sense of, a huge mountain of data. Farr quickly realized that these data could serve a far greater purpose: saving lives. Life expectancy could be broken down and compared over geographic regions, down to the county level. Information about the occupations of deceased persons was also recorded, so Farr could also begin to tabulate life expectancy according to economic and social station. Information about the cause of death was lacking, and Farr probably exceeded his initial authority by adding instructions to list the cause(s) of death on the standard form. This simple addition opened a vast new world of medical statistics and public health that would eventually be called epidemiology, involving the study of patterns of incidence, causes, and control of disease conditions in a population.

In his role as compiler of abstracts Farr issued annual reports to the registrar general on public health and other topics within his purview. In his first report of 1839, he appended a letter that made clear the important uses for the statistics from his office:

> Diseases are more easily prevented than cured, and the first step in their prevention is the discovery of their existing causes. The Registry will show the agency of those causes by numerical facts and measure the intensity of their influence and will collect information on the laws of vitality with the variation in these laws in the two sexes at different ages and the influence of civilisation, occupation, locality, seasons and other physical agencies whether in generating diseases and inducing death or in improving public health.[3]

In composing this grand statement, Farr effectively rewrote his job description. He was not to be simply a "compiler" of facts, but rather he was determined to become an influential advocate for the careful use of data toward the goal of improving the health of the nation.

Over the next forty years, working in the GRO, Farr revolutionized the analysis of the causes of mortality from these records and introduced ideas for identifying risk factors for disease by tabulating deaths from a given disease by various potential causes (poverty, living conditions, environmental factors, and so forth). He also realized that, in comparing various groups (men / women, geographic regions, occupational groups), it was necessary to adjust for differences in age distribution, so he introduced the idea of "standardized mortality rate" to adjust for them.

Like Guerry, Farr had access to a wide range of ancillary data from other government agencies on the social and economic conditions in registration districts of England and Wales (house values, education, poverty rate, and so forth), and he was determined to use the available statistical data to test or compare social explanations of differences in mortality in different classes and to use these findings to press for reform, notably in sanitary conditions. His knowledge, zeal, and command of data would meet its greatest test in the outbreaks of cholera in England.

Cholera

Cholera morbus, commonly known as cholera, is an infection of the intestines now known to be caused by a bacterium, *Vibrio cholera*, and spread by contaminated materials, principally water and human feces. The symptoms are intense vomiting, muscle cramps, diarrhea, and dehydration, most often leading to death if not treated.

The first cholera pandemic began in Bengal, India, in the 1820s, where tens of thousands died, and is said to have reached England in October 1831 on a ship coming from the Baltic states. By 1832, it had spread to London and much of the United Kingdom; more than 55,000 people died in England, Wales, Scotland, and Ireland. It struck France heavily; Irish immigrants to the United States and Canada carried the disease there, and shortly thereafter it arrived in Cuba and Mexico, leaving many thousands dead. By the end of the outbreak in 1837, it had become the greatest worldwide pandemic of the century, and the rapid spread and virulence of the disease attracted widespread attention from people concerned with public health.

Farr and the Miasmatic Theory of Disease

Cholera returned to England in 1848, traveling from India through much of Europe, Russia, and the Middle East. A two-year outbreak again claimed over 15,000 lives in London, 50,000 in the United Kingdom, and over 100,000 worldwide.

During this period, the medical and scientific communities had varying opinions about the factors that caused and spread the disease. In France, it was widely believed to be associated with poverty; in Russia, it was believed to be spread by contagion, but with an unknown mechanism; in the United States, the belief was that cholera had been brought by Irish immigrants.[4]

In England, and London particularly, the most common view was that the disease arose from breathing the foul air around the River Thames, which stemmed from the nearly universal practice of discharging untreated sewage directly into the river. This theory was given a name: *miasma*, from the Greek word meaning "pollution," came to mean poisonous air or vapors filled with particles from rotting organic matter. At this time, London had become the most populous city in the world, and the olfactory evidence of stink near the Thames was undeniable. *Miasma* was a beautiful theory, but in ten years it would be destroyed by data and graphs.

Farr, who by now was statistical superintendent of the GRO, quickly saw that he had all the data necessary to test and compare conflicting theories and explanations of the disease. In 1852 he completed his *Report on the Mortality of Cholera in England, 1848–49*, which the *Lancet* heralded as "one of the most remarkable productions of type and pen in any age or country."[5]

Farr's report is indeed a monument of statistical thinking and visual exploration in the mid-1800s. It comprised over 500 pages and contained 300 pages of tables, charts, diagrams, and maps, following a 100-page introduction describing his analyses and conclusions. Here he described the detailed steps he took to try to test various social and environmental hypotheses about the possible causes of cholera. One can best get a sense of the importance Farr attached to this report, as well as his verbal style to ensure it would get government attention, from his opening paragraph:

> If a Foreign Army had landed on the coast of England, seized all the seaports, sent detachments over the surrounding districts, ravaged the population through summer, after harvest destroyed more than a thousand lives a day for several days in succession, and, in the year it held possession of the country, slain fifty-three thousand two hundred and ninety-three men, women, and children—the task of registering the Dead would be inexpressibly painful; and the pain is not greatly diminished by the circumstance that in the calamity to be described the

minister of destruction was a Pestilence that spread over the face of the island, and found in so many cities quick poisonous matters ready at hand to destroy the inhabitants. (p. i)

As Guerry had done with crime statistics in France, Farr used the mortality records from the GRO and a wealth of other data to examine the distribution of cholera in England over time and space, seeking any regularity of differences due to associated factors linked to the prevalence of the disease. His effort was comprehensive and meticulous. He listed many unusual cases (of one death in Hereford county in 1849, "The common drink of the people is cider," p. li) but went on to examine the possible influence of a myriad of potential factors—environmental (temperature, rain, wind), demographic (age, sex, occupation), and social (poverty, property value, population density)—that might be linked with the prevalence of the disease.

Farr's Diagrams

Figure 4.1 is one of five lithographed plates (three in color) that appear in Farr's report. Farr takes many liberties with the vertical scales (we would now call these graphical sins) to try to show any relation between the daily numbers of deaths from cholera and diarrhea to metrological data on those days. Most apparent are the spikes of cholera deaths in August and September. Temperature was also elevated, but perhaps no more than in the adjacent months. The weather didn't seem to be a sufficient causal factor in 1849. Or was it?

Plate 2 takes a longer view, showing the possible relationship between temperature and mortality for every week over the eleven years from 1840 to 1850. This is a remarkable chart—a new invention in the language of statistical graphs. This graphical form, now called a *radial diagram* (or *windrose*), is ideally suited to showing and comparing several related series of events having a cyclical structure, such as weeks or months of the year or compass directions.

The radial lines in Plate 2 serve as axes for the fifty-two weeks of each year. The outer circles show the average weekly number of deaths (corrected for increase in population) in relation to the mean number of deaths over all years. When these exceed the average, the area is shaded black (excess mortality); they are shaded yellow when they are below the average (salubrity).

Similarly, the inner circles show average weekly temperature against a baseline of the mean temperature (48° F) of the seventy-nine years from 1771

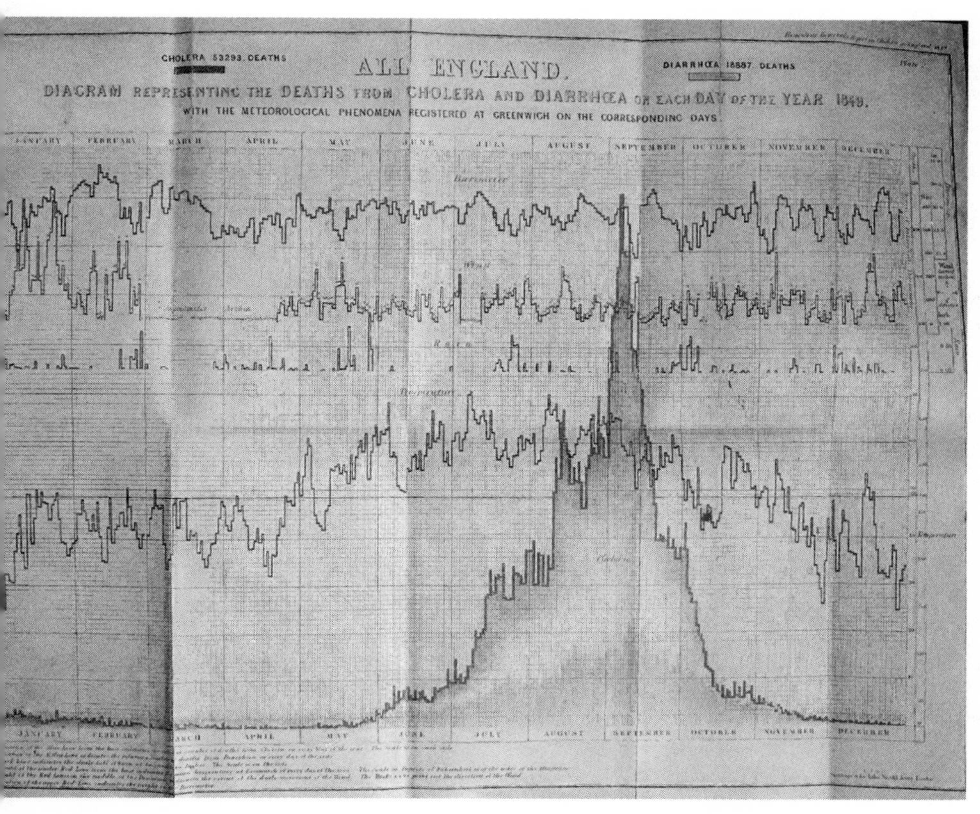

4.1 **Deaths from cholera and diarrhea:** Farr's chart showing deaths from cholera and diarrhea in each day of the year 1849, together with charts of weather phenomena over this time period. The three line graphs at the top show, respectively barometric pressure, wind, rain, and temperature, recorded at Greenwich. The two at the bottom record the numbers of deaths from cholera (dark, with a large peak) and diarrhea (light). *Source:* General Register Office, *Report on the Mortality of Cholera in England, 1848–49*. London: Printed by W. Clowes, for HMSO, 1852, Plate 2.

to 1849. Weeks exceeding this average are outside the baseline circle and shaded red, while those weeks that were colder than average are said to be shaded blue (but appear as gray).

In this graph we can immediately see that something very bad happened in London in summer 1849 (row 3, column 2), leading to a huge spike in deaths from July through September, and the winter months in 1847 (row 2, column 3) also stand out. This larger view, using the idea later called "small multiples" by Tufte,[6] does something more, which might not be noticed in a series of separate charts: it shows a *general pattern* across years of fewer deaths on average in the warmer months of April (at 9:00) through September (at 3:00), but the dramatic spikes point to something huge that can not be explained by temperature.

This radial chart was the immediate precursor to Florence Nightingale's famous "rose diagram" or "coxcomb" showing the causes of mortality among soldiers in the Crimean War (Plate 4). Nightingale is often given credit as the inventor of this graphical form, but she clearly got inspiration from Farr, of whose work on social hygiene she was a great supporter. In fact, the true credit for this graphical invention belongs to André-Michel Guerry,[7] but Farr did this far better than Guerry, and Nightingale corrected a perceptual flaw in Farr's graphic, as we will see later in this chapter. Nonetheless, the diagram by Farr in Plate 2 certainly belongs in the pantheon of the top twenty or so best graphics in the history of data visualization because it tries to show a complex relation between cholera mortality and temperature over time and seasons in a novel graphic form. The link between cholera mortality and weather data did not pan out, so Farr turned to something more directly related to miasma theory—elevation above the Thames as a predictor of cholera mortality.

Farr's Natural Law of Cholera

Farr describes his studies of the distribution of cholera over time and space in enormous detail; and, as Guerry did in his analysis of crime data, he was looking for some sort of law-like behavior in cholera mortality—a combination of stability over factors that don't matter and variation in factors that do. He found this in his analysis of the data from London: "The elevation of the soil in London has a more constant relation with the mortality from cholera than any other known element. The mortality of cholera is in the inverse ratio of the elevation" (Farr, 1852, p. lxi).

Farr found that if he arranged the registration districts in order of their general elevation above the high water mark of the Thames at Trinity, cholera was most lethal in low districts and decreased in inverse relation to elevation above the river. But more importantly, he saw that the relation was tolerably close to a mathematical one, which we would now symbolize as $y \sim 1/x$. He found that, starting with a mortality rate of 1 at 20 feet for example, the mortality rates at successive multiples ("terraces") were closely in proportion to 1/2, 1/3, 1/4, 1/5, 1/6, and so on.

Moreover, Farr found that, given the cholera mortality rate, C, at elevation, E, he could calculate the mortality C' at a higher elevation, E', by the old rule of three from Graunt's political arithmetic, but expressed inversely:

$$E : E' \text{ as } C' : C \quad \Rightarrow \quad C' = \frac{E}{E'} C.$$

Including a constant offset a, to make the scales work, gave the simple formula,

$$C' = \left(\frac{E + a}{E' + a} \right) C.$$

He demonstrated how this worked (with an estimated $a = 12.8$) by calculating the expected mortality at elevations of 0, 10, 30, 50, 70, 90, 100, and up to the highest elevation of 350 in Hampstead. The resulting mortality numbers, 174, 99, 53, 34, 27, 22, 20, and 6 were so close to the actual values (177, 102, 65, 36, 27, 22, 17, 7) that he believed he had discovered their law.

This was precisely the kind of result Farr had been seeking: a definite relation for cholera mortality that could be described by a mathematical formula. But he didn't stop there: to make the relation to elevation visually apparent, he constructed a diagram of these numbers, which is shown in Figure 4.2. In this diagram, it seemed natural to show elevation on the vertical axis, from 0′ at the bottom to 350′ at the top. At each elevation, he drew a horizontal line to show the calculated relative mortality from cholera at the districts of that elevation, and also used dotted lines to show the average *observed* mortality in those districts.

This diagram is also remarkable as an early graph that attempts to show the agreement between data and theory. He includes all the predicted numbers in the body of the diagram, and the dotted lines, representing actual mortality, provide a visual goodness-of-fit test for his theory, showing impressive agreement.

Today, to try to determine if mortality from cholera was systematically related to elevation as an explanatory variable, the natural graph would be a scatterplot of cholera on the vertical axis (y) versus elevation on the horizontal (x). However, although the idea of a scatterplot had been invented by Herschel in 1833 (see the discussion in Chapter 6), Farr and his contemporaries were unaware of it. This idea would only come to the fore in the 1885 work of Galton, thirty-three years later. Therefore, Farr put elevation on

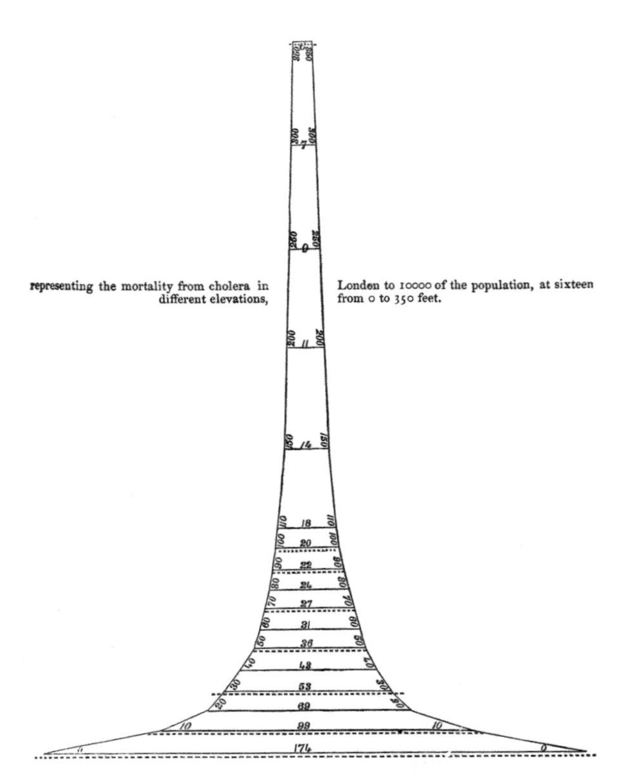

4.2 **Death and elevation:** Diagram of the inverse relation between number of deaths from cholera (width of horizontal lines) and elevation above the Thames (vertical position) for sixteen elevations, from 0 to 350 feet. Numbers on the lines give the predicted values; dotted lines, where present, show the actual number of deaths. *Source:* General Register Office, *Report on the Mortality of Cholera in England, 1848–49*. London: Printed by W. Clowes, for HMSO, 1852. p. lxv.

the vertical axis and mortality on the horizontal, and mirrored the horizontal lines to the left and right to give the visual impression of a tower or monument to his natural law of cholera. It is instructive to reconsider Farr's visual and statistical thinking in terms of what we do currently.

A contemporary graph of Farr's London data[8] for the thirty-eight registration districts is shown in Figure 4.3, with the extreme points labeled. The values calculated from Farr's inverse law of mortality are shown by the solid curve. This curve, with the x and y axes interchanged, gives exactly the right half of Farr's diagram in Figure 4.2. The dashed curve shows a smoothed curve

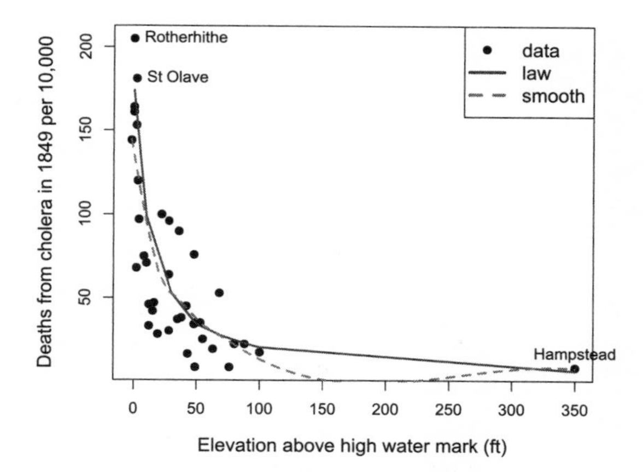

4.3 **Scatterplot:** Farr's elevation—mortality data as a scatterplot, showing mortality (y) as the outcome, in relation to elevation (x) as the explanatory variable. The values of cholera mortality calculated from Farr's "law" are shown by the solid curve. A smoothed fit to the data points is shown by the dashed curve. *Source:* © The Authors.

fit to the data points, tracking the average value of cholera deaths in relation to elevation; it agrees very well with the values calculated from Farr's law. This graph also makes apparent that Hampstead is unusual in elevation and is the reason that Farr's diagram is so tall.

We can gain further insight into Farr's thinking using a different way to examine the relation he found between mortality and elevation. His use of the rule of three to calculate mortality C' for a higher elevation E' from the values of C and E at a lower elevation indicates that he did not think of or understand the more general idea of a statistical relationship of the form $y = f(x)$ for some function f. Had he done so, he might have stated his law as the simpler inverse function,

$$\text{Cholera mortality} = \frac{1}{\text{Elevation} + a}.$$

Doing this yields an extremely simple linear relation between cholera mortality and the reciprocal of elevation. Farr might have called this inverse "lowness" toward the Thames, which was actually what he meant. Figure 4.4 shows the data in Figure 4.3, with elevation transformed to $1/(E + 12.8)$. The theoretical relation he calculated, shown as the solid curve, is now highly linear (the correlation is 0.85).

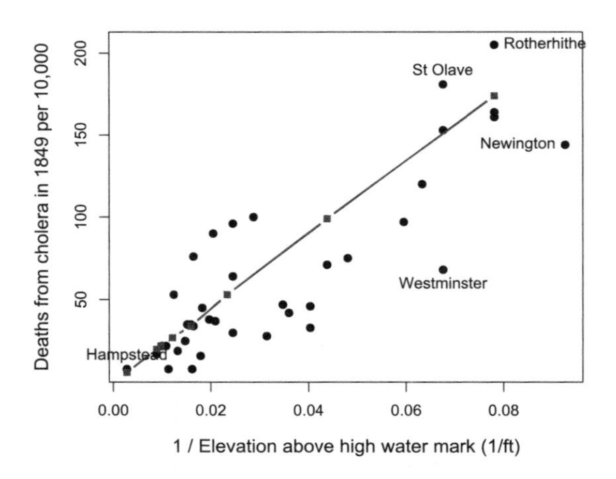

4.4 **Inverse elevation plot:** Plot of Farr's data, with elevation reexpressed as $1/(E + a)$. The predicted values from his theory are now shown to form a highly linear relation to cholera mortality in the solid curve. A few unusual districts are labeled. *Source:* © The Authors.

The Transcendent Effect of Water

Farr was certainly meticulous in evaluating the impact of potential causes on mortality from cholera. But he lacked an effective method for doing so, even for one potential cause, and the idea of accounting for the combination of several causes stretched him to the limit. His general method was to prepare tables of cholera mortality in the districts of London, broken down and averaged over classes of a possible explanatory variable.

For example, Farr divided the 38 districts into the 19 highest and 19 lowest values on other variables and calculated the ratio of cholera mortality for each; elevation had the largest ratio (3:1), while all other variables showed smaller ratios (e.g., 2.1:1 for house values). Having hit on elevation above the Thames as his principal cause, he prepared many other tables showing mortality by districts also in relation to density of the population, value of houses and shops, relief to the poor, and geographical features.

Figure 4.5 illustrates the depth of this inquiry, in an ingenious semigraphic combination of small tables for each district overlaid on a schematic map of their spatial arrangement along the Thames. The tables show the numbers for elevation, cholera deaths, deaths from all causes, and population density, and identify the water companies supplying each district. Unfortunately, this

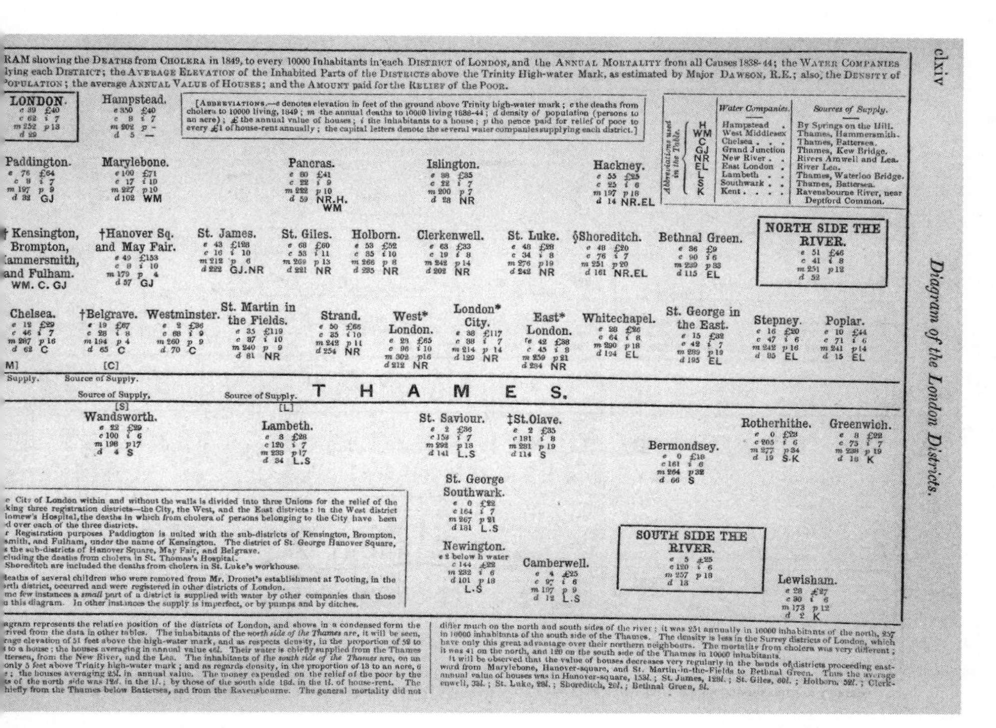

4.5 **Schematic map:** Diagram of the registration districts of London, showing elevation (e), cholera deaths (c), deaths from all causes (m), population density (d), other variables and initials for the water company supplying each district. *Source:* General Register Office, *Report on the Mortality of Cholera in England, 1848–49*. London: Printed by W. Clowes, for HMSO, 1852, p. clxv.

lovely diagram concealed more than it revealed: the signal was there, but the wealth of detail provided too much noise.

It would later turn out that the direct cause of cholera could be traced to contamination of the water supply from which people drew. It was probably confusing that water was provided by nine water companies, as Farr shows in Figure 4.5, so he divided the registration districts into three groups based on the region along the Thames for their water supply: Thames, between Kew and Hammersmith bridges (western London), between Battersea and Waterloo bridges (central London), and districts that obtained their water from tributaries of the Thames (New River, Lea River, and Ravensbourne River).

To see why Farr was misled in his analysis, we replot the data from Figure 4.3 in a new graph at the left of Figure 4.6, coloring the points by water

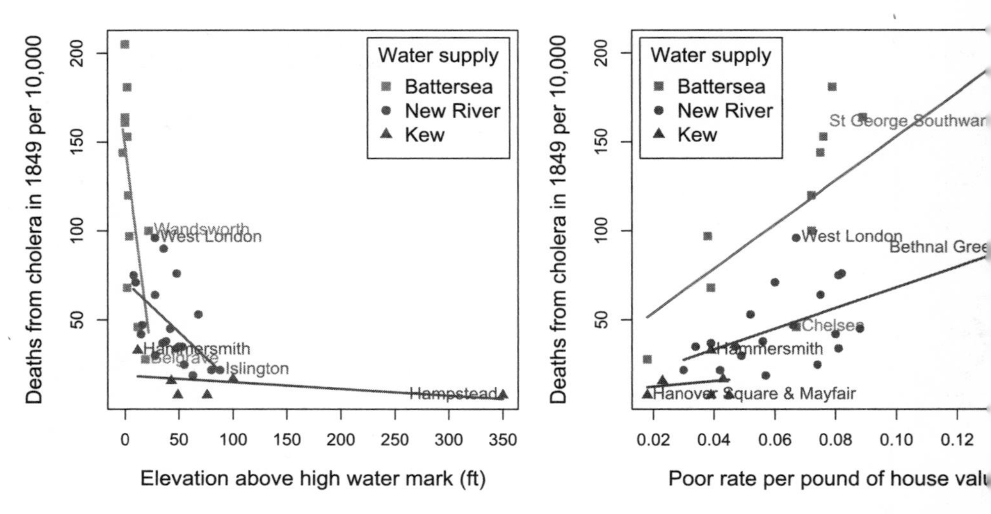

4.6 **Deaths by water supply region:** Reanalysis of Farr's data by water supply region. Left: cholera mortality versus elevation; right: mortality versus poor rate. The lines show linear regression relations for each subset of districts. A few unusual districts are labeled in each plot. *Source:* © The Authors.

supply and drawing a separate regression line for each. It can be seen at a glance that the relation between elevation and cholera mortality differed dramatically for the three water supply regions. Central London (labeled Battersea), which is mostly at low elevations, showed a dramatic decrease in mortality with small increases in elevation. The water there was probably most contaminated closest to the Thames. Western London (labeled Kew) was up river, and so the water was probably less polluted. The slope of the line is still negative, indicating a decrease in mortality with elevation, but only barely so. However, that line probably reflects more the extreme position of Hampstead. The remaining districts (labeled New River) have an intermediate slope for mortality in relation to elevation.

In modern parlance, the relation between elevation and cholera is best described as an *interaction* effect: The effect of elevation on mortality differs substantially for the three water-supply regions.[9] But the average elevation also differed among water supplies, and Farr had no way to see this from his tables and diagrams, which used elevation as the primary cause. Among other modern terms, one could say that water supply acted as a *moderator* variable, a *confounding* variable, or a *lurking* variable.

Farr also explored a variety of other variables as potential causes. He thought that mortality might reasonably increase with population density and poverty (the poor rate) and decrease with the values of houses and shops across London. Once again, his method of analysis concealed the different effects that these variables might have across water-supply regions. The right panel of Figure 4.6 shows the relation of mortality to the poor rate, classified by water supply, with separate lines of fit for each. The slope of the line (the effect of poor rate on mortality) again varies substantially with the water supply.[10]

Farr was certainly brilliant as a compiler and analyst of the statistical data on cholera. But he and most other medical authorities of the day were stuck on the wrong theory: miasma as a cause, and elevation above the Thames as an explanation for its effect. This was a classic example of mistaking correlation for causation: districts at higher elevation had less mortality because the water there tended to be less contaminated.

John Snow on Cholera

Another terrible wave of cholera struck London toward the end of summer 1854, concentrated in the parish of St. James, Westminster (the present-day district of Soho). This time, a correct explanation of the cause would eventually be found with the aid of meticulous data collection, a map of disease incidence, keen medical detective work, and logical reasoning to rule out alternative explanations. It is useful to understand why John Snow succeeded while William Farr did not.

The physician John Snow [1813–1858] lived in the Soho district at the time of this new outbreak. He had been an eighteen-year-old medical assistant in Newcastle upon Tyne in 1831 when cholera first struck there with great loss of life. At the time of the second great epidemic, in 1848–1849, Snow observed the severity of the disease in his district. In 1849, in a two-part paper in the *Medical Gazette and Times*[11] and a longer monograph,[12] he proposed that cholera was transmitted by water rather than through the air and passed from person to person through the intestinal discharges of the sick, either transmitted directly or entering the water supply.

Snow's reasoning was entirely that of a clinician based on the form of pathology of the disease, rather than that of a statistician seeking associations with potential causal factors. Had cholera been an airborne disease, one would expect to see its effects in the lungs and then perhaps spread to others by respiratory discharge. But the disease clearly acted mainly in the gut, causing vomiting, intense diarrhea, and the massive dehydration that led to death. Whatever causal agent was responsible, it must have been something ingested rather than something inhaled.

William Farr was well aware of Snow's theory when he wrote his 1852 report.[13] He described it quite politely but rejected Snow's theory of the pathology of cholera. He could not understand any mechanism whereby something ingested by one individual could be passed to a larger community. To Farr, who was then considered the foremost authority on the outbreak and contagion of cholera, Snow's contention of a single causal agent (some unknown poisonous matter, *materies morbi*) and a limited vector of transmission (water) was too circumscribed, too restrictive. Snow presented his argument and the evidence to support it as if ingestion and waterborne transmission could be the *only* causes; he also lacked the crucial data, either from a natural experiment or from direct knowledge of the water that cholera victims drank.

Farr and others in the medical community found inconsistencies in reports and other reasons to doubt Snow's univocal explanation: "Bad water no doubt sometimes immediately induces the disorder; but we must not suppose it is the sole cause of it."[14] Farr concluded his section on "Bad Water—Dr. Snow's Theory" with a dismissive shrug, saying

> Observations sufficiently exact to decide these questions definitively have yet to be made, and discussed on the principles of probability. The decisive facts cannot be investigated by experiments in which human life may be exposed to risk. They must be carefully looked for and noted by good observers. Conflicting theories serve, among other purposes, to direct the attention of observers to important points which they may otherwise neglect.
>
> While fully admitting the importance of theories, I have endeavoured to present, from the Returns, a view of the facts, without reference to any theory; and to show, independently of the theories, that the conditions in which cholera is or is not fatal, may be determined, and yield important practical deductions.

This comment gives some insight into Farr's view of the role of data as evidence in the search for the cause(s) of cholera: he is a compiler and organizer of statistical facts, not a scientist testing competing explanations against the data. Theories may serve to "direct attention . . . to important points," but the data were primary.

This view was characteristic of most of the people ("data scientists" of their era) who founded the Statistical Society of London (SSL) in 1834. This group later gained a royal charter and was renamed the Royal Statistical Society in 1887. The original logo for the SSL was a bound shaft of wheat, with the motto *ex aliis exterendum* ("for others to thresh out"), meaning, "we gather the wheat; it is for others to make the bread." Farr and others called themselves "statists," not statisticians. The society was concerned only with quantitative facts; interpretation and opinion were to be left to others.[15] Today, the motto of the Royal Statistical Society is more fitting: "Data, Evidence, Decision."

The Broad Street Pump

Snow's opportunity to test his theory came with the new eruption that began toward the end of August in 1854. His celebrated 1855 report, *On the Mode of Communication of Cholera*,[16] describes it dramatically:

> The most terrible outbreak of cholera which ever occurred in this kingdom, is probably that which took place in Broad Street, Golden Square, and the adjoining streets, a few weeks ago. Within two hundred and fifty yards of the spot where Cambridge Street joins Broad Street, there were upwards of five hundred fatal attacks of cholera in ten days. The mortality in this limited area probably equals any that was ever caused in this country, even by the plague; and it was much more sudden, as the greater number of cases terminated in a few hours. (p. 38)

The full story of Snow's discovery of the waterborne cause of cholera has been told in rich detail many times, by medical historians[17] and cartographers,[18] and it was brought to the attention of statisticians and those interested in the history of data visualization by Edward Tufte.[19]

The short, if slightly apocryphal, version of this story is that, during the outbreak of cholera in Soho in 1854, Snow created a dot map of the locations

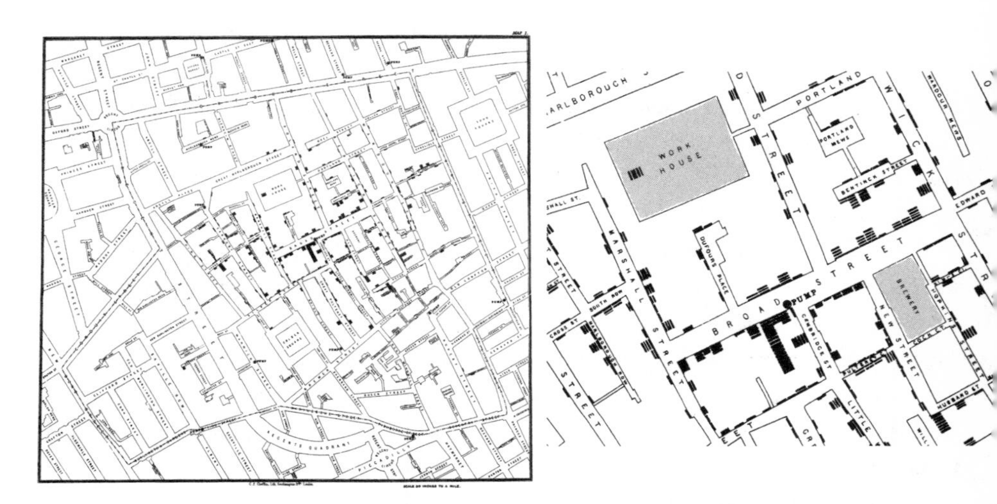

4.7 **Snow's map:** John Snow's dot map of Soho, showing the clusters of cases of cholera in the epidemic of 1854 from August 19 to September 30. Deaths from cholera are shown by stacked black bars at the address of residence. At right: a detail from the map, centered around the pump on Broad Street, highlighting the locations of the workhouse and the brewery. *Source:* John Snow, *On the Mode of Communication of Cholera*, 2nd ed. London: John Churchill, 1855 / Wikimedia Commons.

of deaths and immediately noticed that they clustered on Broad Street, near the site of one of the public pumps from which residents drew their water. This narrative continues: Snow recognized that cases of death were strongly associated with drinking water from this pump. He petitioned the Board of Guardians of St. James Parish to remove the pump handle, whereupon the cholera epidemic subsided.

Tufte was among the first to show that the classical story was not all it seemed (the main apocryphal bit being that the epidemic *ceased* when the pump handle was removed), but that should not detract from an appreciation of Snow's contributions to epidemiology, thematic cartography, and data visualization.

In an effort to show the link between cholera transmission, mortality, and water supply, Snow carried out two studies, both of which he considered to be natural experiments. The first, and most celebrated, was his South London study of cholera mortality in the area of St. James, Westminster, and the most notable result was his now-famous map, shown in Figure 4.7.

In this map, Farr showed the street location of 578 cholera deaths from August 19 to September 30, among the 614 cases whose address could be determined. Deaths were shown by small black bars, which were stacked where more than one victim lived at a given address. The complete map, shown at the left, also shows the locations of the thirteen water pumps that served this district.

This is an example of how an Age of Data contributed to the use and development of data-visualization methods to give new insights into important problems, but it also highlights something new: the difference between aggregate official statistics and individual case-study methods now widely used in epidemiology. By this time, Farr and the GRO had begun recording deaths from cholera on a weekly basis in lists that gave the details recorded from reports in the registration districts. But Snow went further by shoe-leather epidemiology, making visits to houses in the area and interviewing survivors about the sources of their water. He explained,

> On proceeding to the spot, I found that nearly all the deaths had taken place within a short distance of the pump. There were only ten deaths in houses situated decidedly nearer to another street pump. In five of these cases the families of the deceased persons informed me that they always went to the pump in Broad Street, as they preferred the water to that of the pump which was nearer. In three other cases, the deceased were children who went to school near the pump in Broad Street. Two of them were known to drink the water; and the parents of the third think it probable that it did so. (p. 39)

Snow began work on his map sometime in the fall of 1854. Using dots or other marks on maps to show the incidence was not Snow's invention.[20] But he designed his map with a new clarity to make the association of mortality with the Broad Street pump more visually apparent. First, he used as a base map a more schematic form, eliminating all detail except the basic street layout; this may have been based on one prepared by Edmund Cooper, an engineer for the sewer commission, after the end of the outbreak. Second, he stacked the bar symbols to reflect multiple deaths at a given location, making them more visually prominent. Finally, he added icons (•) and "PUMP" labels to show the proximity of deaths to the pumps from which people drew their water. What *was* new in Snow's map was the close connection between the visual display, logical reasoning about the etiology of the disease, and the beginning of a scientific explanation of the mode of transmission.

The contrasting circumstances of those who lived versus those who died were another important feature of Snow's "natural experiment." At some point he recognized two striking anomalies in the area near the pump (highlighted in the detail at the right of the •). He wrote in his report:

> The Workhouse in Poland Street is more than three-fourths surrounded by houses in which deaths from cholera occurred, yet out of five hundred and thirty-five inmates only five died of cholera, the other deaths which took place being those of persons admitted after they were attacked. The workhouse has a pump-well on the premises, in addition to the supply from the Grand Junction Water Works, and the inmates never sent to Broad Street for water. (p. 42)

Similarly, he noticed that there were no deaths among the more than seventy brewer's men working at the local brewery. Snow called upon the proprietor, who informed him that the brewery had its own deep well on the premises. Moreover, "The men are allowed a certain quantity of malt liquor, and Mr. Huggins believes they do not drink water at all; and he is quite certain that the workmen never obtained water from the pump in the street." It is hard to resist echoing Tufte's comment: "Saved by the beer!"[21]

The Neighborhoods Map

The version of Snow's map shown in Figure 4.7 is the most famous, but a second version is more interesting graphically and scientifically. The Cholera Inquiry Committee, appointed by the Vestry of St. James, submitted its report on July 25, 1855.[22] The section titled "Dr. Snow's Report" contained a new map that attempted a more detailed and direct visual analysis of the association of death with the Broad Street pump.

This new map, shown in Figure 4.8, states and tests a geospatial hypothesis: people are most likely to draw their water from the nearest pump (by walking distance). The outlined region in this map "shews the various points which have been found by careful measurement to be at an equal distance by the nearest road from the pump in Broad Street and the surrounding pumps."[23]

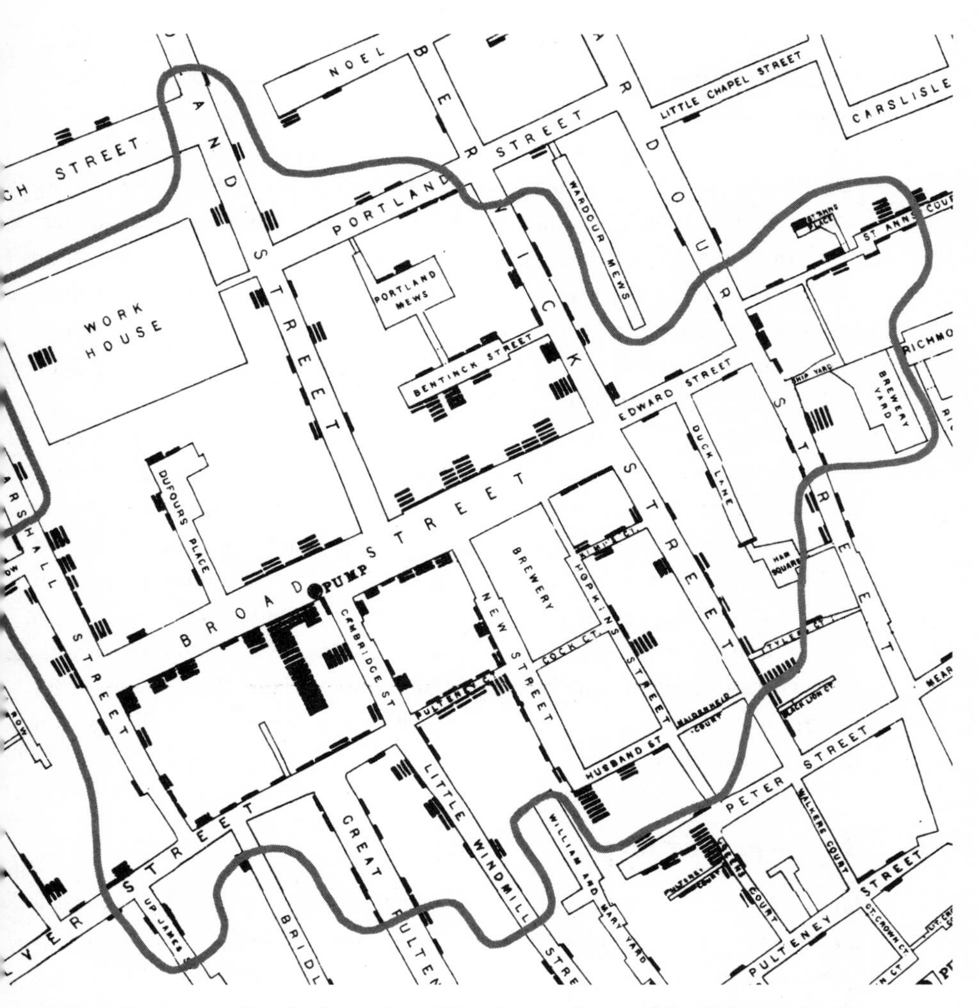

4.8 **Boundary region:** Detail of a version of Snow's second map of the cholera outbreak, showing the boundary of the region of addresses believed to draw their water from the Broad Street pump. *Source:* The John Snow Archive and Research Companion.

If the source of the outbreak was indeed the Broad Street pump, one should expect to find the highest concentration of deaths within this area, and also a low prevalence outside it. He stated his conclusion as "it will be observed that the deaths either very much diminish, or cease altogether, at every point where it becomes decidedly nearer to send to another pump than to the one in Broad street."[24]

The final explanation for the source of the outbreak came slightly later, through the work of Reverend Henry Whitehead, the curate at a local church and a member of the Cholera Inquiry Committee. He identified the first ("index") case as the death of a five-month-old infant, Frances Lewis, whose family lived at 40 Broad Street, immediately adjacent to the pump. When severe diarrhea struck the child, her mother, Sarah Lewis, soaked the diapers and emptied the pails into the cesspool at the front of their house, only three feet from the well. Unfortunately, the cesspool walls had decayed and the effluent flowed directly into the pump well. Thomas Lewis, the baby's father and a local constable, suffered a fatal attack of the disease on September 8, the same day that the pump handle was removed.

John Snow died in June 1858, before his waterborne theory was taken seriously by the medical community. In 1866, a new cholera outbreak occurred in Bromley, East London. William Farr still did not accept Snow's waterborne explanation as the sole cause of transmission; but from new data, he realized that it was at least credible enough to issue a "boil-water advisory" through the GRO. The discovery of the causative agent, the bacillus *Vibrio cholerae*, is now credited to the Italian scientist Filippo Pacini [1812-1883], who examined with a microscope the intestinal mucosa of patients who had died from cholera in an 1854 outbreak in Florence and saw a comma-shaped organism that he called *Vibrio*.

Re-visioning the Broad Street Pump

Snow's data and his map have become such classics in the lore of epidemiology and thematic cartography, that many people have attempted to reproduce or "improve" his map, in various ways, and for various purposes; these have not always been either historically accurate or with a positive effect.[25] Two revisions of Snow's map follow, both of which attempt answers to the question, "How could Snow have made his map more visually effective for his purpose?" but with different presentation goals and audiences in mind.

Figure 4.9 is a very simplified (or dumbed-down) version in the style of a presentation graphic that Snow might have used in a PowerPoint presentation to the Board of Guardians in his petition to remove the pump handle (but we're fairly certain Snow would have rejected this). It is actually two steps removed from Snow's original. In a 1958 paper titled *Pioneer Maps of Health and Disease in England*,[26] the Oxford social geographer Edmund William

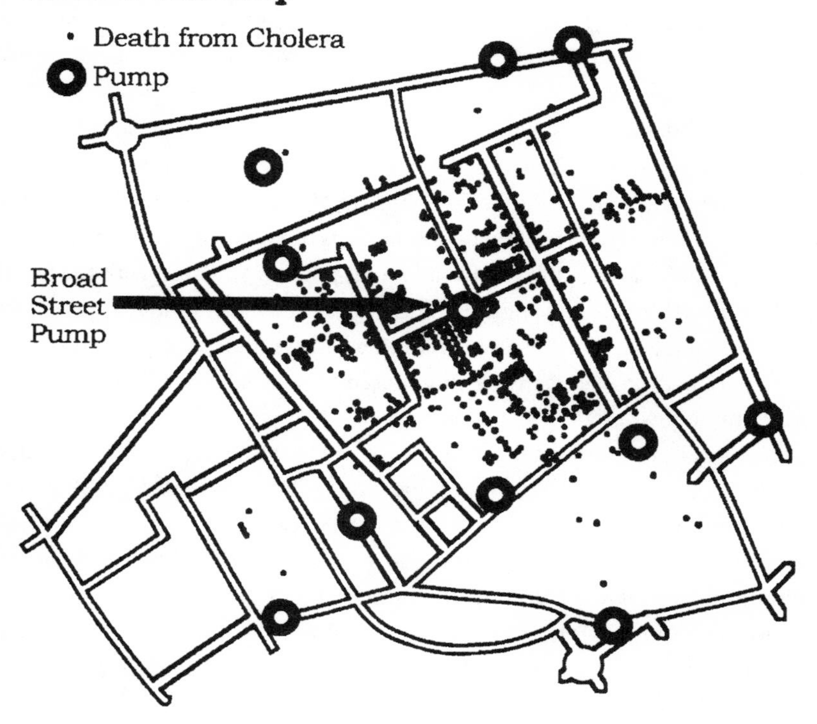

4.9 **Presentation graphic:** Mark Monmonier's re-vision of the Gilbert version of Snow's map, as a presentation graphic. *Source:* Extracted from Mark Monmonier, *How to Lie with Maps*. Chicago: The University of Chicago Press, 1991, fig. 9.18.

Gilbert drafted a slightly simpler version of Snow's map, retaining only the major street names and replacing the black bars for deaths with dots. He also removed the labels of the anomalous cases of the workhouse and brewery that were crucial to Snow's argument. Gilbert carelessly captioned his version, "Dr. John Snow's map (1855) of deaths from cholera . . ." and misled later authors in thinking that this was indeed Snow's map.

The re-vision in Figure 4.9 was pared down from Gilbert even more by Mark Monmonier in *How to Lie with Maps*. He removed all place names, made the dots for deaths slightly smaller, and greatly magnified the circle symbols used for the pumps, adding a big arrow pointing to the one on Broad Street. About the only thing he didn't do to the map was to use the title "BROAD STREET PUMP CAUSES CHOLERA" in large bold type. One could argue that this was acceptable if the presentation goal was consent by the Board of Guardians to remove the pump handle. Lost in this translation, however, is Snow's attempt to show visually the relation between cholera mortality and the sources of water.

From a different perspective, Plate 3 shows two modern statistical re-visions of the central portion of Snow's map, using data sets carefully digitized from Snow's original.[27] Effective graphics answer the question: *Compared to what?* What Snow really wanted people to see is the very high concentration of deaths in the area closest to the Broad Street pump, compared with the lower levels of deaths near other pumps. This re-vision overlays the map with two graphical enhancements ("layers" in modern geographic information system [GIS] parlance) to help achieve this presentation goal—graphs that convey Snow's message directly to the eyes.

The lines in the top panel of Plate 3 are the boundaries of what are called Voronoi polygons around the pump locations. They answer the question, for each pump, what region in the map is closest to that pump, relative to all other pumps?[28] The shading levels show the high concentration of deaths surrounding the Broad Street pump compared to other pump regions.

For each pump region, we tallied the number of deaths.[29] If there was no association between pump and deaths, they would be uniformly distributed throughout the map and the expected number of deaths would be proportional to the area of each region. The result of this analysis is shown in Table 4.1. The column labeled "Difference" shows the difference between the observed and expected frequencies. The polygons in this panel of Plate 3 are shaded in proportion to the value of the standardized difference (residual) using red for positive values; it is clear that there were far more deaths than expected around the Broad Street pump.

In Snow's time, there were no statistical methods to test the hypothesis that deaths in the pump regions differed from chance expectation. The χ^2 test was only developed in 1900 by Karl Pearson. Such a test confirms the overwhelming numerical (Table 4.1) and visual evidence (Plate 3) that cholera deaths were disproportionally concentrated in the area of the Broad Street pump.

As a second enhancement, the contour lines in the bottom panel of Plate 3 show the smoothed average intensity of cholera deaths calculated at a fine grid of points across the (x, y) positions in the map. They are like the contours

Table 4.1. Frequency analysis of deaths by pump neighborhoods. The expected frequencies show the values that would have occurred if deaths were uniformly distributed over the pump neighborhoods according to their areas. The Broad St. pump stands out for a huge excess of cholera deaths.

Pump	Actual # of Deaths	Expected # of Deaths	Difference	Standardized Residual
Broad St	359	49	**310**	44.3
So Soho	64	34	30	5.1
Crown Chapel	61	72	−11	−1.3
Warwick	16	40	−24	−3.8
Briddle St	27	28	−1	−0.2
Oxford St #2	24	55	−31	−4.2
Oxford St #1	12	25	−13	−2.6
Gt Marlborough	6	48	−42	−6.0
Vigo St	4	84	−80	−8.7
Coventry St	2	53	−51	−7.0
Dean St	2	44	−42	−6.4
Castle St E	1	11	−10	−3.1
Oxford Market	0	35	−35	−5.9

of elevation in topographical maps and can be thought of as the outlines of horizontal slices at increasing height, showing hills and valleys in the terrain of cholera deaths. They are also shaded (from light green to intense red in a color version) to draw attention to the highest-density region, which nearly coincides with the Broad Street pump.

If we imagine ourselves as present-day statistical consultants to John Snow, he most likely would have wanted more: an interactive map that could be zoomed in or out to show more or less detail; certainly the ability to click on a point to show a text box with the circumstances of the victim; and perhaps the ability to add other layers that summarized or illustrated the effect of other factors. Nonetheless, he probably would have been well pleased with these revisions.[30]

Graphical Successes and Failures

As the compiler of abstracts, Farr was more inclined to use tables than graphs to present his reports to the home secretary and parliamentary committees. Yet the scale and importance of the cholera outbreaks over many years gave him so much data on so many variables that he was led to the use of charts to try to show patterns and seek relations with cholera mortality.

By this time, Playfair's line graphs of time-series data (see Chapter 5) were relatively well-known and Farr's use of this device in Figure 4.1 was an attempt to determine whether deaths were related to weather phenomena over time. This was certainly novel in the application of multivariable time-series graphs to disease mortality, and he is likely the first to have introduced this idea in the areas of public health and epidemiology.

The prevailing theory of miasma or airborne transmission certainly responded to the direct sensory evidence of the stink of effluent discharged directly into the Thames. Farr thought that he had found the link in the strong inverse relation of mortality to elevation (Figure 4.2). As we have seen, however, he was misled by the confounding relation to water supply (Figure 4.6), and he failed to see this because of a limited graphic vision.

Like Playfair's time-series charts, Farr's were essentially what we call "1.5-dimensional" (1.5D)—something between a univariate graph and a fully 2D bivariate graph. He could understand plotting *X* (temperature, . . .) and *Y* (mortality) versus time, but not the idea of plotting *Y* versus *X* directly, no less the idea of trying to assess the direction or strength of such relations. This was all awaiting the invention of the scatterplot and the measure of correlation, which would come later (Chapter 6).

If Farr's role is sometimes underappreciated, John Snow's role and legacy in this story of data visualization are sometimes overappreciated, although both were important, if for different reasons. Snow was certainly correct in attributing the spread of cholera to a water-borne agent, and he was also correct in his visual demonstration of this as concentrated around the Broad Street pump in the famous dot map (Figure 4.7) of the 1854 cholera outbreak in the Soho district. But, as Tom Koch, a modern critic has observed,

> Science isn't about being right. It is about convincing others of the correctness of an idea through a methodology all will accept using data everyone can trust. New ideas take time to be accepted because they compete with others that have already passed the test.[31]

Snow's problem was that he was so certain that cholera could *only* be transmitted by contaminated water that he failed to deal convincingly with other possible explanatory factors: elevation, soil conditions, housing density, and so forth. To gain acceptance, a new theory, like a new king, must often decisively dispose of other contenders to the throne. Farr, and others in the medical community were used to dealing with multiple causes, now called "risk factors" in epidemiology, and they gnashed their teeth in public debate, criticizing Snow.[32]

The Answer: A Bug

Eventually, Snow's hypothesis was proved to be correct, but only long after his death in 1858. The causative agent, the bacillus *Vibrio cholerae*, was discovered with a microscope by the Italian scientist Filippo Pacini in 1854. But this finding seems to have passed largely unnoticed. The very idea that a microscopic living organism could be the cause of the disease was revolutionary and nearly unfathomable.

It wasn't until a new cholera outbreak occurred in east London in 1866 that William Farr presented more compelling statistical evidence that this outbreak had been caused by sewage-contaminated water; but, like Snow, he had no idea of an organism-based explanation of the mechanism. It remained for the German physician Robert Koch [1843–1910] to isolate the bacillus in a pure culture in 1884, and show that the organism was always found in patients with cholera but never in those with similar symptoms (diarrhea) from other causes.

Scientific and historical appreciations often undergo mood swings. Koch justifiably received the Nobel Prize in Physiology in 1905 for his contributions, which also included the discovery of the tubercle bacillus, the main causative agent of tuberculosis. But it took until 1965 for an international committee on nomenclature to adopt *Vibrio cholerae Pacini 1854* as the correct name of the cholera-causing organism.

Much later, the pendulum finally swung back to an appreciation of John Snow as the guy who got it right with the aid of a graph. Such stories, even if somewhat apocryphal, still serve a purpose. They help us to understand the connections among the hard work of data tabulation and summary, and then the effort to turn data into insightful graphical displays. But in science, it is always necessary to sell your idea to contemporaries persuasively. And nothing counts as much as a correct causal explanation that eliminates the alternatives.

Florence Nightingale's Graphical Success

If William Farr's beautiful radial diagram (Plate 2) had no impact because he was displaying the wrong variables, another graphical contribution by Florence Nightingale [1820–1910] to vital statistics in this period changed health policy forever. Moreover, it corrected what is now considered a blunder in the graphic portrayal of counts of deaths by Farr in this figure.

Florence Nightingale, who is widely known as the mother of modern nursing, is called "the lady with the lamp." She was also a social reformer with a keen understanding of the power of graphics for persuasion, and consequently was also called a "passionate statistician."[33]

Nightingale was born to a wealthy, landed British family. As a young girl, she exhibited an interest in and flair for mathematics, which was encouraged by her father, William. Later, she was profoundly influenced by reading Adolphe Quetelet's 1835 *Sur L'Homme et le Developpement de ses Facultés*, in which he outlined his conception of statistical method as applied to the life of man.[34] She also felt a strong religious calling to the service of others, and against her mother's strenuous objections, she decided that nursing would be her vocation.

The Crimean War, which was fought by Russia and the forces of France, Britain, and the remnants of the Ottoman Empire, began in October 1853 and lasted until February 1856. In October 1854, Nightingale appealed to her friend Sidney Herbert, secretary of state for war, to send her and a team of nurses to the Crimea. She soon recognized that most of the deaths occurred, not from battle, but from preventable causes: zymotic diseases (mainly cholera) and insufficient sanitary policy in the hospitals that treated the soldiers.

Nightingale was so appalled by what she witnessed in the Crimea that when she returned to England, she launched a campaign to persuade the British government to create higher standards for the treatment of soldiers in field hospitals.[35] Having kept meticulous records of the causes of mortality, she sought advice from William Farr on how to analyze and present her data.

Unlike Farr, who was an accomplished presenter of statistical "facts," Nightingale's motivation was persuasion—a call to action for the British government to adopt reforms to the entire treatment of soldiers at war, from their diet to the design of hospitals with good sanitary conditions, even considering accommodations for soldiers' wives.[36] She was impressed enough with Farr's

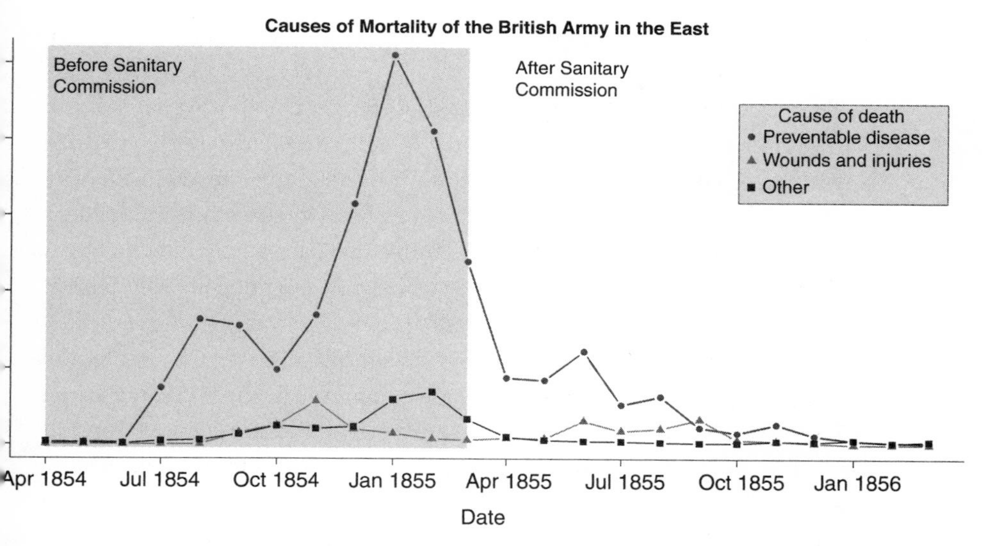

4.10 **Line graph of Nightingale's data:** The data on causes of mortality from Figure 4.12 plotted as a time-series line graph. *Source:* © The Authors.

use of a radial diagram to adopt this form for her own data. In her first version, printed privately for the secretary of war in 1858, she followed Farr's design, which plotted deaths on a linear scale as distances from the origin.[37]

Nightingale quickly realized, however, that this design was deceptive because the eye tends to perceive the area rather than length in such displays: doubling the death rate would give a perceived area four times as large. In her next version, shown in Plate 4, Nightingale plotted deaths in each month as the square roots of distance from the center, so the area of each wedge reflected the number of deaths. Here it is easily seen that deaths from preventable diseases such as cholera (the outer blue edges) totally dominate those from battlefield wounds and other causes.

Nightingale's visualization showed[38] that deaths during the first seven months of the Crimean campaign amounted to an annual rate of 60 percent per annum from disease alone—a rate that exceeded that of the Great Plague in London in 1665–1666 and that of the British cholera epidemics in 1848 and 1854. Following her persistent requests to the War Office, a sanitary commission was formed around April 1855 to investigate the causes of high mortality of the British Army in the Crimea. A series of reforms were instituted, and as seen in the left panel of Plate 4, deaths from preventable causes began to quickly diminish, dropping to a mere 4 per 1,000 by March 1856.

The result in Plate 4 is not quite as dramatic as the apocryphal story of Snow removing the handle of the Broad Street pump, but it was something anyone could clearly see, and Nightingale's "rose diagram" entered history as a powerfully persuasive graphic depiction of medical intervention and sanitary practice. As Stephen Stigler stated, "It is ironic that Farr, who plotted in hopes of discovery, produced a misleading picture, while Nightingale, who plotted for rhetorical purposes, did not mislead."[39]

Had Nightingale been influenced by William Playfair (see Chapter 5), she might have drawn a time-series line graph comparing deaths by months from the same three causes, as shown in Figure 4.10.[40] The data are the same, but this graph doesn't have quite the same graphical impact factor as Nightingale's. She wanted her message to hit the viewers between the eyes. Her rose diagram succeeded, changing medical practice in war and peace forever.

5

The Big Bang

William Playfair, the Father of Modern Graphics

The Big Bang theory is the current model of cosmology for the origins and subsequent evolution of the physical universe. Its most fundamental characteristic is that over a time span no longer than the blink of an eye the universe went from almost nothing to almost everything. This is an apt metaphor for the sudden development of statistical graphics from rudimentary forms, largely poorly executed, to polished displays indistinguishable from the best that have been done since.

This cataclysm occurred toward the end of the nineteenth century, when nearly all the modern forms of data graphics—the pie chart, the line graph of a time series, and the bar chart—were invented, and the key developments were due to a wily Scot named William Playfair. He can rightly be called the father of modern graphical methods, and it is only a slight stretch to consider his contributions to be the Big Bang of data graphics.

Playfair's legacy and his place in the history of data graphics stem largely from two main works. The first, *The Commercial and Political Atlas*, was first printed in a privately circulated form in 1785, followed by progressively extended editions in 1786 and 1787, a French translation in 1789, and a final third edition in 1801. In this atlas, he introduced the idea of line graphs to compare economic data on imports and exports to England from various countries over time and the bar chart for data that were not time-based. His second main work, *The Statistical Breviary* (1801), presented comparative statistical data on a variety of topics for the countries of Europe; here he introduced the use of the pie chart to show proportions of a whole.[1]

Playfair's Life

William Playfair was born on September 22, 1759, in the small Scottish village of Liff, near the city of Dundee. His father, James, a minister in Liff, died in 1772, and so the responsibility for twelve-year-old William's education fell to his older brother, John, who, even at the young age of twenty-four, was already recognized as likely to become one of Scotland's most distinguished natural scientists.

John's approach to science was unequivocally empirical. Later William recalled an assignment given to him by his brother in which he was to record the daily high temperatures over an extended period of time. John told him to think of his results as a series of thermometers sitting side by side and to record them graphically with that form in mind. It was but a small step to abstract the essence from the images of the thermometers. He would record a dot, representing the top of the column of mercury in the thermometer, as an appropriate point in a Cartesian space: the time of the measurement as the horizontal axis and the temperature on the vertical axis. William subsequently gave credit to his brother for instilling the idea of translating numerical information into a spatial form. As he said later,

> The advantages proposed by [the graphical] mode of representation, are to facilitate the attainment of information, and aid the memory in retaining it: which two points form the principal business in what we call learning. Of all the senses, the eye gives the liveliest and most accurate idea of whatever is susceptible of being represented to it; and when proportion between different quantities is the object, then the eye has an incalculable superiority. (*Breviary*, 1801, p. 14)

His "two points in what we call learning" was one of the earliest expressions of the idea that the impact of a graph depends on both how features of data are represented ("encoded") and how they are understood and remembered in the mind of the viewer. He understood that

> the eye is the best judge of proportion, being able to estimate it with more quickness and accuracy than any other of our organs . . . this mode of representation gives a simple, accurate, and permanent idea, by giving

form and shape to a number of separate ideas, which are otherwise abstract and unconnected. (*Atlas*, 1801, p. x)

In this, he was among the first to express explicitly the connections between eye, brain, and understanding derived from graphs. He also anticipated modern ideas in cognitive psychology such as depth of processing by noting that people remember information better when they process it in a meaningful rather than superficial way: "Information that is imperfectly acquired, is generally as imperfectly retained."

William started on this path in 1774, at the age of fourteen, when he left the family home to apprentice with Andrew Meikle, a well-known Scottish engineer and the inventor of an early threshing machine. Three years later, William was recommended to the position of draftsman and assistant to James Watt, during the early days of the Birmingham steam engine factory, where Watt was beginning to develop automatic recording devices for temperature and pressure. The lessons he learned as a draftsman would serve him well in the future when he turned his hand to writing. His interests expanded beyond engineering and manufacturing to issues of economics, and he realized that the methods of graphical display, which he found so useful in seeing the structure of temperature changes, had even greater potential to clarify the murk of economic data now becoming broadly available.

In 1785 Playfair prepared a preliminary edition of his now iconic *Commercial and Political Atlas* and circulated it for suggestions. James Watt, who at that time was justly famous for perfecting the steam engine, which revolutionized manufacturing, was skeptical of the graphs and charts of economic data featured in the *Atlas*. His principal concern, still manifest today, was the lack of authority represented by the figure—data shown in graphs don't look as serious as those in tables. Watt believed that although someone could make up figures out of whole cloth, they would not make up numbers, an assumption that history has repeatedly shown false. Watt was also concerned by the lack of precision of the representation.

Playfair agreed that a tabular representation could show data to greater precision, but he argued that often such accuracy was unnecessary and was unrelated to the goals of the representation. Although he stated this repeatedly, he was never clearer than in the introduction to the *Atlas*:

> The advantage proposed by this method, is not that of giving a more accurate statement than by figures, but it is to give a more simple and permanent idea of the gradual progress and comparative amounts, at different periods, by presenting to the eye a figure, the proportions of which correspond with the amount of the sums intended to be expressed. (*Atlas*, 1801, ix–x)

He subsequently (in the introduction to the *Breviary*) made the modern argument that the graphical presentation is *more* truthful because it does not show the data to greater accuracy than is deserved:

> With respect to throwing aside the units, tens and hundreds, in great numbers, it is done under this simple impression, that as the information does scarcely come within a thousand of the truth, it is an affectation of accuracy beyond what has really been attained; or, to make a fair comparison, it is like a historian giving as truth, an account of the private minutiae of courts and embassies, which were known only to the parties themselves, and though reported publicly never believed. No sort of reflection is however meant on those who think fit to give their statements in the other way, although the number of figures certainly embarrasses the memory without answering any good purpose. (*Breviary*, 1801, 6–7)

It is in these ideas: the eye as a better judge of comparisons than a table of numbers and even simple line graphs as a way to make trends and patterns more understandable and memorable, that Playfair revolutionized the essential idea of what we now call a "graph."

Playfair's Graphic Contributions

Over the course of almost forty years William Playfair made a large number of key contributions to effective graphic display of data. His key innovations were time-series line graphs, plotting some variable over time; an extension of this, shading the area between two such curves to show their difference; the bar chart, showing a quantitative variable as a rectangular bar along some scale, with different bars for different circumstances to be compared; and

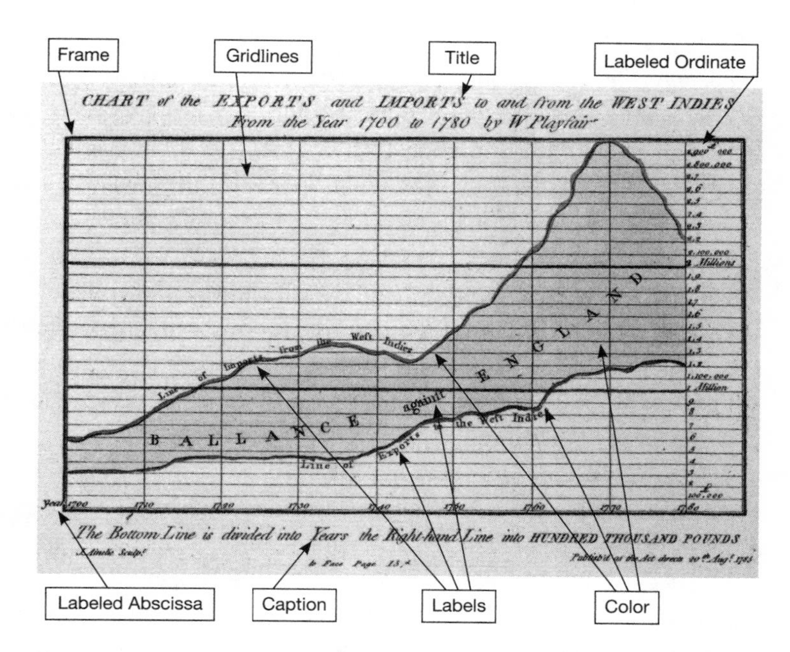

5.1 **Graphical conventions:** Some of the basic graphical conventions for statistical charts established by Playfair. These features were designed to make graphs more directly readable and understandable. *Source:* William Playfair, *The Commercial and Political Atlas*, London, 1786. Labels by the authors.

the traditional pie chart and variations on the same theme, sometimes called "circular diagrams," to show proportions of a whole.

Various graphic design elements had been used before, principally in maps. In his graphic works, Playfair can be seen to have developed some elements of a graphic language for charts of data that are now considered standard conventions in published data graphics; some of these are shown in Figure 5.1. Among the embellishments and refinements to graphs that he formulated and improved were the following.

1. *Framing* of the plot, leaving room inside for labels and axis values.
2. *Titles*, placed outside the frame or as a framed cartouche inside, and describing what was shown or the purpose of the chart.
3. *Color coding*; for example, he used a thick red line for exports, a green line for imports, and he filled in the space between them with

one color when exports exceeded imports (the balance of trade was positive) and a contrasting color when the balance of trade was negative. At the time, this required hand-coloring of printed copies, but Playfair thought this was important for graphic communication.

4. *Hachure and stippled dots*, used when color was either too expensive or not available; Playfair simulated dark colors by hachure and lighter colors by stippling.
5. *Labeling of axes*, including the name of the variable and units.
6. *Gridlines*. Major gridlines were engraved more heavily than minor gridlines, giving two levels of scale.[2]
7. *Time-period indicators*, highlighting spans of time that serve to help understand trends or differences.
8. *Suppression* of nonsignificant digits. Playfair often rounded axis labels and data values shown to make them more easily seen.
9. *Event markers* locating historical events in time. He instigated a practice, too little followed today, of placing vertical gridlines at important dates, even if this meant spacing them unequally. We illustrate this in Figure 5.6 with Playfair's plot of the soaring national debt of England, one of the first-ever rhetorical plots to try to tell a graphic story with historical context.
10. *Theoretical, hypothetical, or projected values* and the use of solid and broken lines to represent them—typically using solid lines to show what is observed and broken lines to show what is theoretical, hypothetical, or projected.

Not all these visual conventions were novel, but Playfair's great contribution was the way that he mobilized all these elements in combination to enrich both the display and the information it conveyed to the viewer. In this, he created the essential ideas of modern graphs designed to communicate quantitative facts in the context of a narrative.

The First Pie

Circular diagrams of various kinds going back to antiquity were used to represent periodic phenomena of time (a clock, signs of the zodiac) and space (the Ptolemaic and Copernican systems). As noted by the American Pie Council, the first pie recipe goes back to the Romans (a rye-crusted goat cheese and honey pie), and it was natural to cut it into wedges. But it awaited Playfair to adapt the sliced pie to data display.

The modern pie chart made its first appearance in the *Statistical Breviary* in 1801. Playfair's goal here was to show, "on a principle entirely new, the resources of every state and kingdom in Europe." In his first chart, shown in Plate 5, he wanted to show the overall size of the countries and the subdivisions within them, as well as other features designed to tell a graphic story of their prosperity. He did this using a sequence of circles, where the area of each circle represents the geographical size of the country before the French Revolution of 1789.

Labels within these circles show the composition of the Russian and Turkish Empires by continent. All the rest are solely European. The countries stained brown are maritime powers; the pale red ones are only powerful by land. The yellow portion of the Turkish Empire indicated the small segment that lay on the African continent. The two concentric circles for the Russian Empire showed that it occupied parts of two continents, mostly in Asia, but was still the largest country in Europe. Next in size is the Turkish Empire, which is situated on all three continents. Playfair originally tried representing this as three concentric circles but was unhappy with the accuracy with which a viewer would perceive their relative sizes, so instead he represented the three areas as three sectors of the same circle. Thus was born the first pie chart.

Playfair typically designed data-rich plots, and this was no exception. In addition to the physical size of each country and its continental location, inscribed over each circle is a number (e.g., 14 for Sweden) that indicates the number of people per square mile of that country; the numbers inside the circles show the country's area in square miles. Finally, the red line on the left of each country shows the number of inhabitants (in millions), and the yellow line on the right of each country represents its revenue in millions of pounds sterling. By combining data on population, revenues, and surface area, he created what can be considered an ancestor of multivariate data displays.

Playfair explained that the dotted lines connecting the population and the revenue "are merely intended to connect together the lines belonging to the same country. The ascent of those lines being from right to left (e.g., Portugal, Britain & Ireland and Spain) or from left to right (all the others), shows

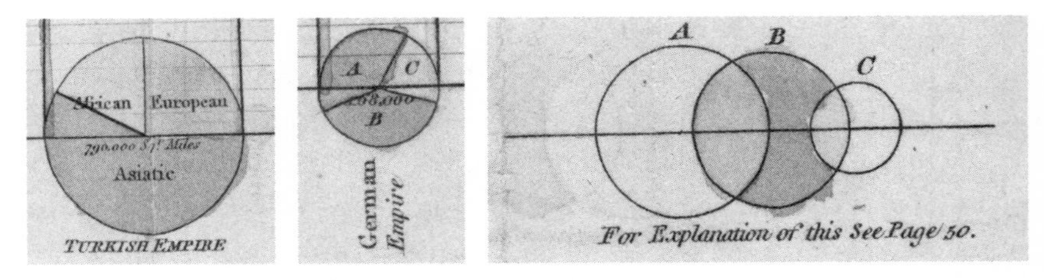

5.2 **Pie chart details:** Three details from Playfair's Chart 2, showing two pie charts (left: Turkish Empire; middle: German Empire) and a Venn-like diagram (right) for the German Empire. *Source:* William Playfair, *The Statistical Breviary*, London, 1801.

whether in proportion to its population the country is burdened with heavy taxes or otherwise." These tilted lines invite the visual inference that their slope reflects the relative burden of taxes per person, but this would be wrong because (a) the nation circles have varying diameters and (b) population and revenues are shown on different scales. Nevertheless, the sloped lines do distinguish Spain, Portugal, and Britain and Ireland (lines tilting up) from the others (lines tilting down), and this is probably all that Playfair intended.

Playfair included a second circle and pie chart in the *Breviary* (Chart 2), showing similar data on extent, population, and revenues of European countries in 1801, after the division of Poland and the Treaty of Luneville between France and Austria, which reflected gains by Napoleon and other political realignments. The former German Empire had been divided among Austria (A) and Prussia (C), and the rest as under the control of German princes (B). Figure 5.2 shows the pie chart–relevant details from this chart, which is otherwise similar to Plate 5.

The pie chart for Turkey at the left of Figure 5.2 is similar to that in Chart 1. But the German Empire created explanatory problems that Playfair tried to solve visually: how to compare the old German Empire with the rest of Europe, how to show the new subdivision of political control, and how to show the changes. The middle panel shows the pie chart he substituted for the German Empire, divided by area of political control.

In the right panel of this figure, he attempted something that, to our knowledge, had never been done before in a visual representation: to show overlapping sets in a diagram, with areas proportional to the sizes of the sets and what they had in common by areas of overlap. In this diagram, the leftmost circle (A) shows the size of the new Austrian interests; the rightmost circle (C) shows the size of the Prussian interests. The middle circle (B) represents the former German Empire, whose area is the same in the original as that of the pie chart in the middle panel. The intersection of the areas A and B is supposed to show the area ruled by Austria, and the intersection of B and C, the area under Prussian rule. The remainder of B in the middle represents the provinces of the other German princes.

The abstract idea of logic diagrams to show overlapping sets had originated in medieval times with the Catalan philosopher Ramon Llull [1232–1316] in his 1305 *Ars Magna*. Eighty years after Playfair, John Venn (1880)[3] proposed the formal use of overlapping circles to represent propositions in logic ("All Swedes have blond hair," "Some blondes have more fun") and the logical conclusions that can be drawn from these. This graphical form of intersecting circles came to be known as a *Venn diagram*, and is widely used today.

Playfair's attempt was not entirely accurate graphically, but he was the first to attempt this with data showing the sizes of intersecting sets approximately proportional to the areas of regions. The theoretical problem of how to do this accurately for four or more sets, so that all intersections are shown, is an active topic in computer graphics and computational geometry.

Humble Pie

The principal value of the pie chart is that it can show that all component pieces sum to the total. However, experimental evidence has repeatedly shown that dot plots allow us to see the relative size of each category more accurately and can do so with many more categories than a pie.[4] Dot plots do not show the total, but that seems a small price to pay for their many advantages. It isn't surprising that, after inventing pies for a very specific purpose, Playfair rarely used this format again. So why have pies remained popular for more than two centuries, often finding their way into corporate reports and slick news magazines?[5]

One obvious answer is that filling the slices with color gives a wider range of choices than can be achieved with points and lines. Another feature is their compact size and ease of comprehension (when the number of categories is modest) makes them easy to use as elements of a more complex display. Throughout his work, Playfair used graphical displays to convey complex, data-heavy stories to his readers or viewers. Other masterful adopters of this mode of communication had the same desire, and some were able to take Playfair's idea of the pie chart to a higher level.

In 1858 Charles Joseph Minard [1781–1870] published a map of France (Plate 6) with more than 40 pie charts embedded in it. The area of each circle was proportional to the amount of meat exported by that department to Paris. Each circle was broken down into three categories of meat: colored black for beef, red for pork, and green for lamb. Minard also created the convention of using a different background color for the départements far from Paris, for which the data on meat sent to Paris was either missing or zero.

Minard's map shows something more subtle. Because it is generally more expensive to transport goods over larger distances, one would expect the sizes of the circles to decrease with distance from Paris. But that is not uniformly true in this figure. A huge amount of beef comes from the region east of Paris, but given the distance, the amounts from south-central France, including the Limousin region around the Massif Central, stand out somewhat. Limousine beef continues to be a favorite throughout France and elsewhere.

Thus the humble pie, when combined with other graphic forms (here a map) can quickly convey a complex story. At a glance we see that fewer than a dozen departments supply most of Paris's meat; only four provide most of its pork and three most of its lamb. What makes this an effective graphic display is the use of multiple divided circles distributed geographically on the map of France.[6] The individual pies combine two visual features: their area, representing the total meat, and the angular subdivision by type. Their combination with the map is compelling.

This use of the divided circles (and other subdivided forms) as small multiples in a larger display would later be put to great use in several sets of national statistical albums produced toward the end of the nineteenth century. Among the most impressive is the collection of the *Album de Statistique Graphique* published by France's Ministry of Public Works from 1879 to 1899 under the direction of Émile Cheysson [1836–1910]; these works are described in detail in Chapter 7.

Plate 7 is one fine example among many that used proportional and divided circles to portray various aspects of French commerce. This one is meant to show simultaneously the transport of various types of goods to the ports of Paris by trains or boats arriving on the various canals and rivers, together with the composition of goods arriving at the main maritime ports, Le Havre, Bordeaux, and Marseille. The central pie in Paris shows the total for the city.

Seven categories of goods (plus "other") are shown by the colored pie sectors, including construction materials (red), fertilizers (blue), and "combustible materials" (i.e., coal: black). The bar charts at the bottom right give a further breakdown of these goods by mode of transport and by country of origin.

Time-Series Line Graphs

Playfair's principal graphic contribution, the one he used first and most often, was of a line graph, usually showing economic data over time. He used multiple curves to contrast different circumstances and then to reason about important economic issues that had not previously been demonstrated visually. In so doing, he provided the public and policy makers with a new way to think about those issues.

Playfair's *Commercial and Political Atlas*, which was largely composed of time-series line graphs, aimed to emphasize the balance of trade between England and other trading partners. He understood two centuries ago what has become common knowledge now—that there is more to be gained by adopting a small set of display formats to be used and reused with different data sets than to try to be creative and adopt a novel format for each data set because in this way the reader need only confront an unfamiliar design once.

Playfair's *Atlas* follows this dictum to perfection. Almost all chapters have the same form—a single graph shows the imports and exports between England and the country that is the subject of the chapter, followed by his interpretative narrative discussing what was shown in the graph. Two examples illustrate Playfair's graphical method, but with different narrative goals.

Imports and Exports

Figure 5.3 is typical of Playfair's approach, dealing with trade between England and Germany over the entire eighteenth century. It follows the graphical conventions illustrated in Figure 5.1, with some small amendments: The name for the vertical axis (Money) is shown on the left boundary, and that for the horizontal axis (Time) is shown on the top. In each year, exports exceeded imports, so the area between the two curves is labeled "Balance in Favor of England."

Playfair uses the information in the plot as a jumping-off point for discussion. The data themselves answer the question, "What's happening?" but give

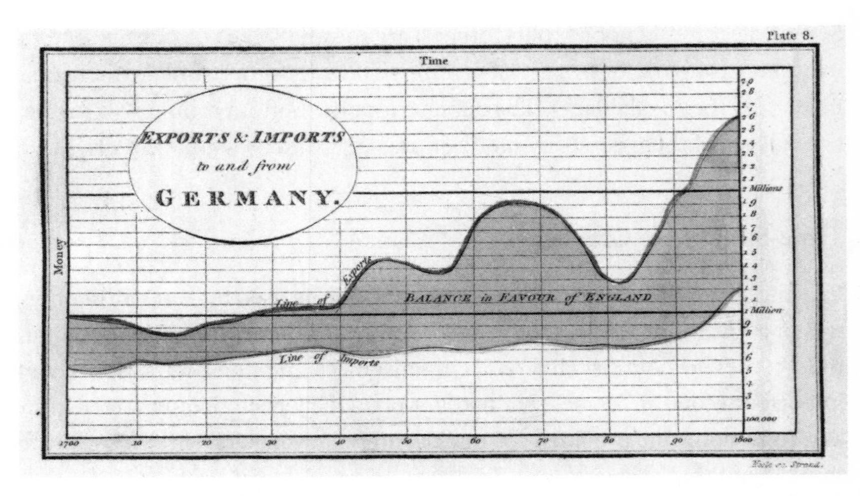

5.3 **Exports and imports with Germany:** A time-series line graph showing England's imports from and exports to Germany over the entire eighteenth century. The area between the curves represents the balance of trade, labeled "Balance in Favor of England." *Source:* William Playfair, *The Commercial and Political Atlas*, London, 1786.

no explanation. Playfair suggests the difference can be understood in terms of English manufacturing:

> The trade with Germany, very considerable in its amount, is also from its nature one of the most advantageous branches of our commerce. . . . Those that we import thence are chiefly raw materials, and our exports consist principally of finished goods, the value of which is derived from the labour and art in making; so that they afford a greater advantage, and are a source of more riches to us than twice the trade might be, if the articles were of a highly intrinsic value.

However, he goes on to explain that "the articles exported to Germany are chiefly of the sort that the Germans manufacture themselves," but that German manufacturing has been hampered by "strict laws made, relating to freedom and corporations" (p. 37). Evidently, Playfair would find himself aligned with some of the conservative, antiregulation forces of today!

A modern reader might look at this graph and note features that call for further explanation: the level of imports from Germany remained relatively constant over a large time span, rising modestly in the last decade. However,

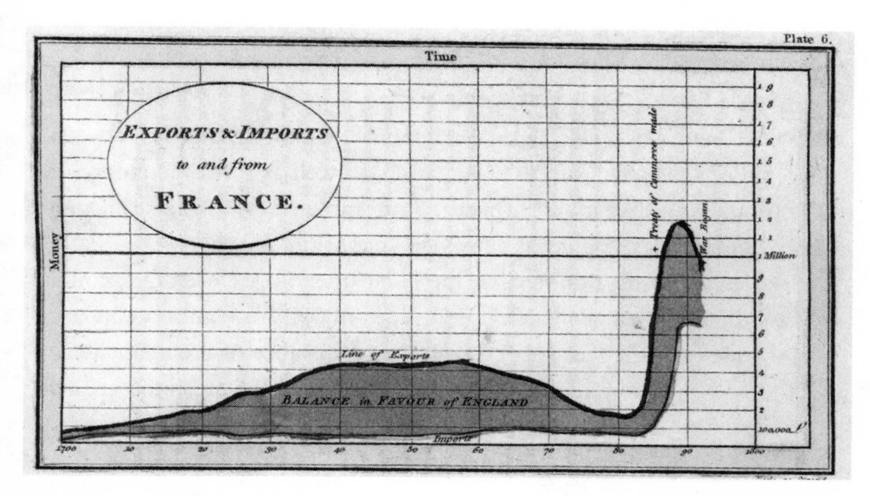

5.4 **Exports and imports with France:** A time-series line graph showing England's exports to and imports from France over the entire eighteenth century. The area between the curves represents the balance of trade, labeled "Balance in Favor of England." *Source:* William Playfair, *The Commercial and Political Atlas*, London, 1786.

exports from England, while rising overall, had several periods of growth and decline, followed by a large increase in the last two decades. This was the power of even such simple line graphs for Playfair. He would have agreed with John W. Tukey that "the greatest value of a picture is when it forces us to notice what we never expected to see."[7]

Playfair knew far too much to accept data on their face; when he spotted what he took to be some sort of distortion, he would use both the graph and words describing it to illuminate the truth, as he understood it. Nowhere is this clearer than in his comments on England's trade with France; also, he was not always gentle in his assessments.[8]

Playfair chose to plot the trade between England and France (see Figure 5.4) on a scale ranging from 0 to 2 million pounds / year. This seems oddly generous: except for the period between 1783 and 1788, when there is a sudden rise, the trade was always less than a half-million pounds / year. Why did Playfair not follow his usual practice of allowing the range of the data to determine the graph's bounds?

Upon seeing it this way, the viewer is immediately struck by how limited the trade was. This is especially true for France and England, which, for reasons of geography and history, ought to have had substantial mercantile relations (the average trade with Germany was ten times that with France). The graph raises the question "what's up?" and Playfair explains:

> We have before us a very fallacious representation of the trade between two nations, which, from their situation, as well as from the nature of the productions, we might expect to find immense; yet which, through a strange species of policy, is extremely inconsiderable.
>
> There cannot be a doubt that the illicit trade far exceeds in amount what is here delineated, which can include only what is regularly entered. This trade furnishes us with an astonishing instance of the inefficiency of the laws that are injudiciously enacted, and which furnish too great a reward for evasion. (p. 31)

Playfair's wisdom was summarized a century later by the English industrialist and statistician Sir Josiah Charles Stamp [1880–1941], who pointed out (in what is sometimes called "Stamp's Law of Statistics") that,

> The government [is] extremely fond of amassing great quantities of statistics. These are raised to the nth degree, their cube roots are extracted, and the results are arranged into elaborate and impressive displays. What must be kept ever in mind, however, is that in every case, the figures are first put down by a village watchman, and he puts down anything he damn well pleases.[9]

Figure 5.4 contains the beginning of another innovation—the insertion of possible causal agents as labels, giving historical context. Upon seeing the sudden increase in trade (both exports and imports) in 1783 the viewer naturally wonders "why?"

Playfair inserted one plausible cause: a Treaty of Commerce around 1785. He also reminded viewers that the sudden downturn in trade in 1789 coincided with the start of the French Revolution—an unlikely time for trade (or the careful, formal, recording of trade) to flourish. These annotations help us understand the reasons behind the effects we see. However, we must visually interpolate between the decennial time labels below the horizontal axis to estimate when the instigating events occurred. Playfair addressed this shortcoming in subsequent designs.

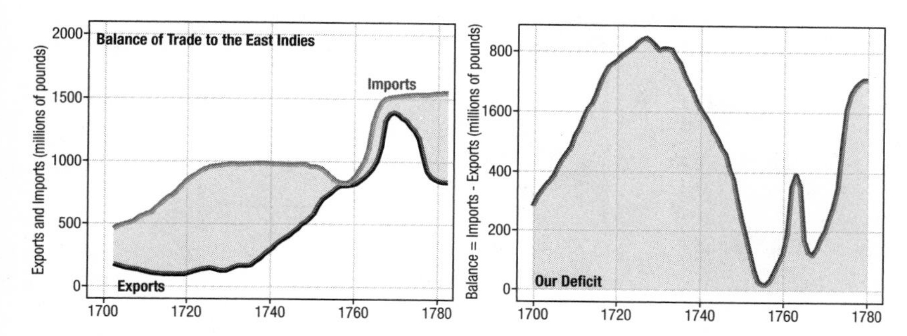

5.5 **Plotting the difference between curves:** Left: Re-creation of Playfair's chart of the exports and imports between England and the East Indies; right: a direct plot of balance of trade, exports minus imports. *Source:* Redrawn from William S. Cleveland, *The Elements of Graphing Data*. New Jersey: Hobart Press, 1994, fig. 4.2.

Playfair's Failure: Problems with Curve-Difference Charts

Playfair seemed to have an intuitive sense of graphic design, and he extolled the virtues of the graphical method to facilitate both seeing and remembering trends in numerical data: "The eye has an incalculable superiority."

His main goal in the series of charts in the *Atlas*, represented here by Figures 5.3 and 5.4, was to show the balance of trade for or against England by the difference or area between the two curves. For this goal, Playfair was less than successful when viewed through a modern lens, but neither he nor anyone else noticed.

Unfortunately, human perception is easily misled in this situation. In a series of experiments in 1984, William Cleveland and Robert McGill demonstrated that the eye cannot reliably perceive the vertical *difference* between two curves and used one of Playfair's charts as an example.

The left panel of Figure 5.5 shows a simple version of the graph of trade between England and the East Indies that appeared in the first edition of the *Commercial and Political Atlas.*[10] It is easy to see that England always imported more than it exported over this period, but the *size* of the trade deficit is hard to see.

The right panel of this graph plots the trade deficit—the difference between the two curves—directly over time. It is surprising to see here that: (a) the trade deficit was greatest around 1726, (b) the period 1755–1770 had a very

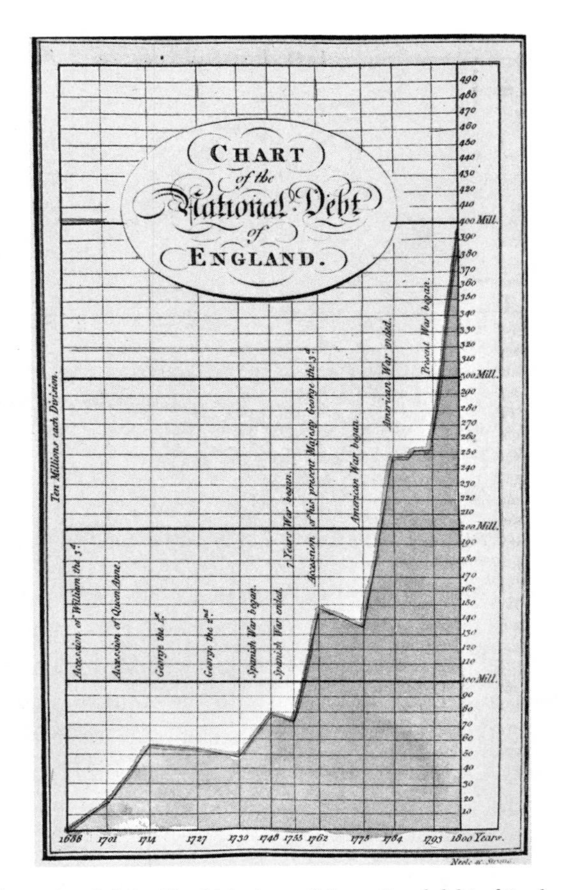

5.6 **Plotting the national debt:** Playfair's chart of the national debt of England containing a lot of explanatory collateral information brilliantly integrated. *Source:* William Playfair, *The Commercial and Political Atlas*, London, 1786.

peculiar wiggle, and (c) the average deficit between 1710–1740 was about as great as that in 1780.

England's National Debt

Figure 5.6 is one of Playfair's most remarkable displays. It contains a mixture of data and design innovations to provide a convincing visual argument connecting data and history. Here, the goal is to connect the national debt of England with significant events, principally wars.

The design begins with the recognition that the grid of a plot serves the same purpose as the scaffolding that workers erect to construct a building; and just as the scaffolding is removed after the building is completed, so too with the graphical gridlines. Playfair recognized that what happened in 1700, 1710, or 1720 is not of primary interest; instead interest on specific dates would arise in two kinds of situations, connecting causes or explanations to historical events.

- When the outcome one sees in the data suggests a question, such as what happened in 1730 or in 1775 to make the debt suddenly rise? Or in 1748 or 1784, when the debt fell or stabilized?
- When a plausibly impactful event occurs and we want to know its consequences for the national debt, such as did the accession of Queen Anne in 1701 or that of George II in 1727 have any effect?

With this idea in mind, Playfair built the graph shown in Figure 5.6. But he replaced the initial, vertical rules for years as decades with gridlines at historically interesting times.

Consequently, the vertical gridlines were no longer spaced equally across the century, and each vertical line was annotated with a plausible causal event. This is a relatively small change, but the effect was to create a stronger link between a graph and a narrative argument.

We see that national debt increased in 1730, when the Spanish War began, and decreased in 1748, when it ended. The debt rose markedly in 1755 at the beginning of the Seven Years War and decreased in 1762 at its conclusion (and the accession of George III). It then jumped from 1775, when the American Revolution began, and abated in 1784, when the war ended. The message is clear: wars are bad for national debt.

Playfair obviously understood that perception of the data is easily manipulated by the ratio of the width of the graph to its height—its aspect ratio. The choice of this ratio is largely subjective. If the horizontal axis is much larger than the vertical scale, a soaring mountain can be transformed into a gradual plain. To make his point about the soaring national debt more emphatically, Playfair chose to make this plot higher than it was wide, a format that probably challenged book production of the day.

Modern research on effective design reveals that the choice of aspect ratio is most often best resolved by selecting a ratio that makes the main pattern in the plot as close to a diagonal line as possible. Playfair's prescient design adhered to this advice, even though the experimental evidence for it lay almost 200 years in the future.

Bar Charts

In preparing the *Atlas*, Playfair was disappointed in the paucity of trade data available between Scotland and its commercial partners. Instead of the century of longitudinal data that existed between England and its partners, he found data on Scotland's trade with England for just one year, 1781. Such data did not allow him to examine trends over time, but such a lack was less serious than would be omitting any discussion of Scotland's trade. In his own words,

> The limits of this work do not admit of representing the trade of Scotland for a series of years, which, in order to understand the affairs of that country, would be necessary to do. Yet, though they cannot be represented at full length, it would be highly blamable entirely to omit the concerns of so considerable a portion of this kingdom.

But how was he to represent Scotland's imports and exports to each of the other nations for only a single year? He invented (or adapted) the bar chart, using side-by-side bars to compare imports and exports (see Figure 5.7), and in the process showed something entirely new.

He said of his bar chart: "This Chart is different from the others in principle, as it does not comprehend any portion of time, and it is much inferior in utility to those that do; for though it gives the extent of the different branches of trade, it does not compare the same branch of commerce with itself at different periods." (*Atlas*, p. 101)

Yet, although Playfair denigrated this invention, his graphic design got it absolutely right (or close enough) the first published time:

- He drew this as a horizontal bar chart, so that the labels for the countries could be written horizontally at the right.
- He grouped the bars by country rather than by imports versus exports. Doing the reverse would have made comparisons of trading partners more difficult.

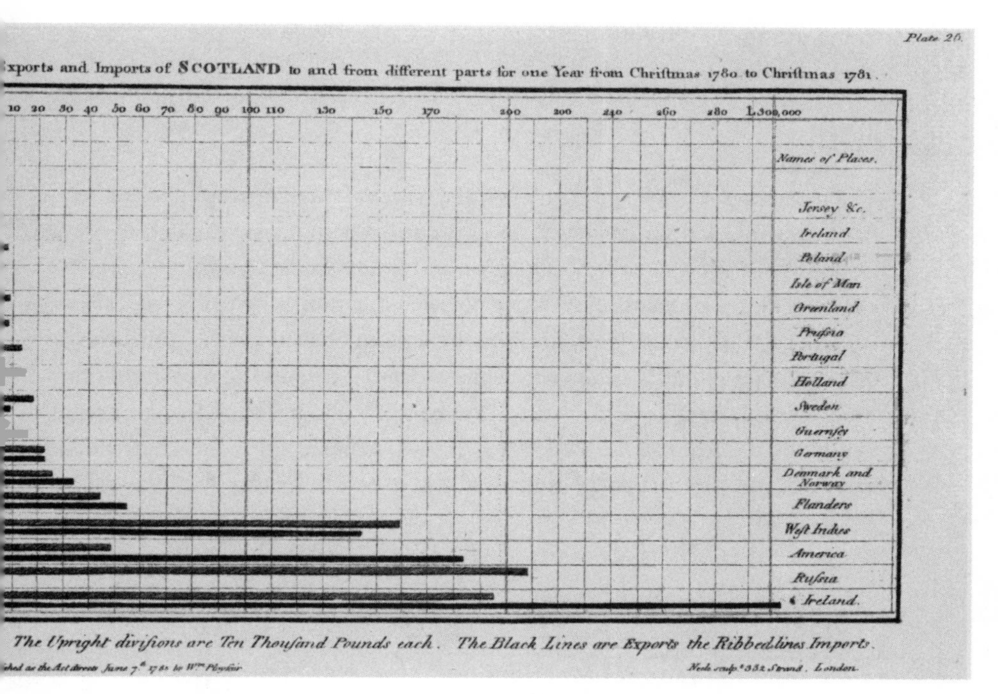

5.7 **Playfair's first bar chart:** Playfair's bar chart of the imports (gray bars) and exports (black) to Scotland in 1781 from 17 different places. *Source:* William Playfair, *The Commercial and Political Atlas*, London, 1786.

- Pairing imports and exports allows them to be directly compared for each of Scotland's trading partners.
- Most importantly, he ordered the places of trade numerically according to the *data* to show the importance of trading partners. Ordering *alphabetically* might have made lookup easier ("Where is Holland?"), but this would have made the overall pattern harder to see.

In this chart, the eye is immediately drawn to the bottom bars, showing trade with Ireland, Russia, America, and the West Indies most prominently. That was Playfair's main message. A closer look shows something more: trade between Scotland and Ireland was dominated by Scotland's exports, in a ratio of 3:2; trade with Russia was nearly entirely imports; with America, Scotland exported more than it imported; with the West Indies, trade was roughly equal. Another convention used here is to explicitly show zeros as "0" rather than an empty bar.

Earlier Bar Charts

Playfair was not the first to conceive of a bar chart: a data graphic using rectangles, whose length conveyed an amount, sometimes subdivided to show portions of the total. As far back as around 1480, Nicole Oresme (Figure 2.2) conceived of the idea to show some physical magnitude by using "pipes."

A decade before Playfair, Philippe Buache published (with Guillaume de L'Isle) *Cartes et tables de la géographie physique*, containing a bar chart and associated data table of high and low water levels in the Seine over time, semiannually from 1732 to 1766. Figure 5.8 is the first true bar chart we are aware of. So why is Buache not usually credited in history for this graphic innovation?

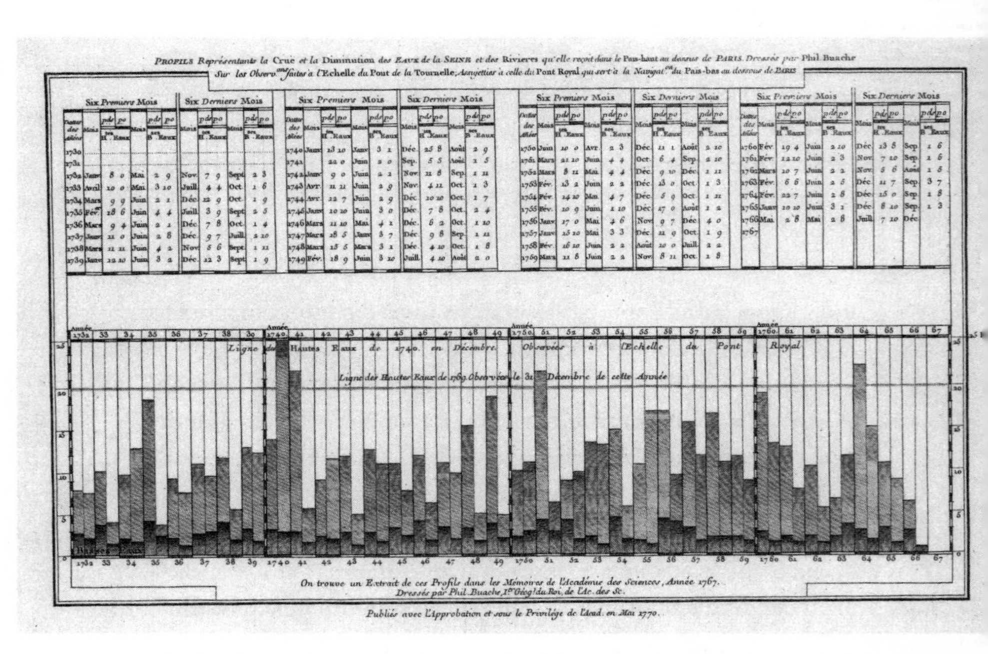

5.8 **The first known bar chart:** A bar chart by Philippe Buache and Guillaume de L'Isle showing both the low and high water marks of the Seine for the thirty-five years between 1732 until 1766. *Source:* Philippe Buache, *Cartes et tables de la géographie physique ou naturelle*, Paris, 1754.

Buache, a physical geographer, was quite used to profile maps of terrain elevation over space. To show changes over time, he substituted time for space, and he used two levels of shading to distinguish high and low water levels. Buache probably didn't recognize that he had done anything novel: he was just making a visual inscription of the table of numbers shown at the top of Figure 5.8, derived from the physical marks of water levels recorded on the Seine.

On the other hand, Playfair developed the idea of using the length of bars to show something intangible: imports and exports to Scotland that could be understood in a new way.

Charts of History

Playfair characteristically took a long, historical view of the data that he reported in his charts. He was not satisfied with his first bar chart (Figure 5.7), for, despite its thoughtful design and execution, it lacked the depth and richness of information that were his hallmark. Because most of his time-series plots included data that spanned a century, he must have felt the limitations of including imports and exports for only one year. He was forced to do this by the lack of additional information, but he didn't have to like it.

Forty years later, in a pamphlet titled *Letter on Our Agricultural Distresses*, Playfair found a use for bars as an enhancement of his beloved time-series plots and published the wonderfully rich plot shown here in Figure 5.9. The bars on the plot show the price of a quarter-bushel of wheat, in shillings, on the east axis. Superimposed below those bars is a curve that depicts the weekly wages of a good mechanic, with values in shillings given on the west axis. The plot spans the 265 years from 1565 until 1830 and is augmented, on the top border, with the names of the reigning monarchs, from Elizabeth I to George IV.

Playfair's principal goal in producing this plot was to provide evidence that, over the more than two centuries depicted, wheat was becoming increasingly affordable for the worker, and also to provide some context for how this relationship changed over historical periods. To do this, he combined three graphic forms: a line graph for wages, which changed relatively smoothly over time; a bar chart of price of wheat, which was more variable over the five-year intervals shown, and the historical timeline at the top showing the reigns of monarchs—all plotted over time.

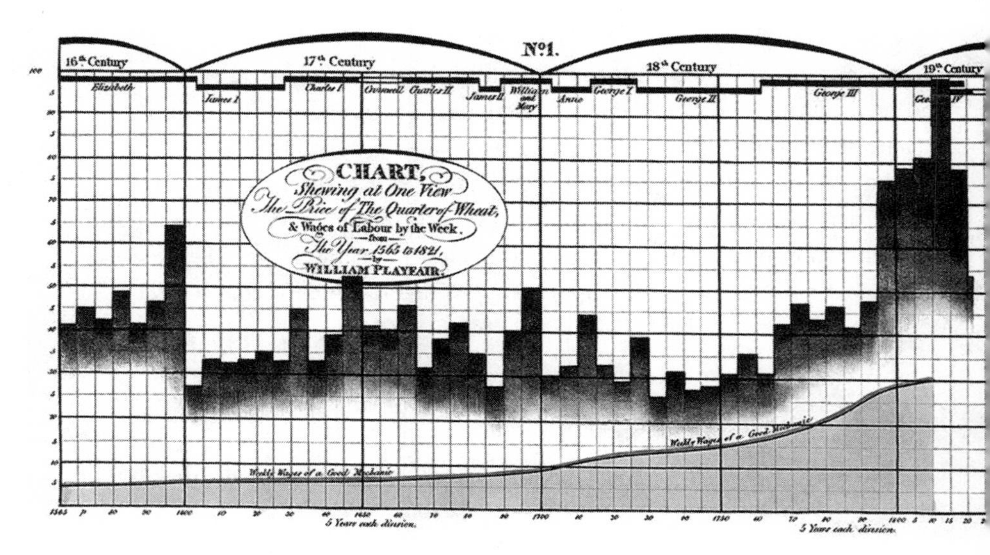

5.9 **Parallel time-series:** A time-series display showing three parallel time-series: prices of a quarter of wheat (the histogram bars), wages of a good mechanic (the line beneath it) and the reigns of English monarchs from Elizabeth I to George IV (1565 through 1820). *Source:* William Playfair, *A Letter on Our Agricultural Distresses, Their Causes and Remedies.* London: W. Sams, 1821.

Yet Playfair's conclusion is not immediately obvious to most viewers. At best, we might consider this chart a lovely infographic, beautifully executed by a master craftsman, and a model for others seeking inspiration in graphic design. But at worst, this chart can be considered a graphic failure, because it does not convey his message directly to the eye. Playfair's problem in Figure 5.9 was that he thought of his plots exclusively in terms of showing variables separately over time and was unable or unwilling to think of plotting a *derived* variable directly. We discuss the plot in Figure 6.5 in Chapter 6 (color version in Plate 10).

One more quite remarkable but less well-known chart by Playfair also deserves mention here. In his 1805 work, *An Inquiry into the Permanent Causes of the Decline and Fall of Powerful and Wealthy Nations,* Playfair surveys the entire history of ancient and modern states from 1500 BCE to the present in his "Chart of Universal Commercial History," shown in Plate 8. Here, the horizontal axis shows time over this range, with vertical divisions and labels marking "remarkable events relative to commerce." The wealth and commerce of ancient states (from Egypt at the bottom to Constantinople at the top) are shown in the bottom portion by filled area charts, with a pink background; the "places that have flourished in modern times" (Spain up to Russia, and then the United States) appear at the top.

The scientist and natural philosopher Joseph Priestley [1733–1804], who was the codiscoverer of oxygen, had introduced the first modern timeline[11] in his 1765 "Chart of Biography," which showed the lifespans of two thousand famous people from 1200 BCE to 1750. A second and more ambitious "New Chart of History" attempted to show the entire history of countries, peoples, and empires over this period.[12]

Playfair clearly knew of Priestley's timeline charts, but what he did in his chart of "commercial history" was something much simpler, more powerful and accessible to visual inspection. He shows each political entity as if it were a distribution of some quantitative measure of wealth and commerce.[13]

Thus, one can easily see the rise and fall of nations and their span of influence over time. His visual message is crystal clear: to maintain wealth and prosperity, nations must attend to commerce, their balance of trade, and national debt.

The visual form Playfair created here, of tiny charts without details to show overall trends and allow direct visual comparison, was made crisper and more generally useful by Tufte in 1983, and these charts were later dubbed "sparklines" (small, intense word-sized graphics). They are tiny, minimalist graphics that can be used as graphic elements in tables or larger charts as Playfair did here.

They can even be embedded in text, to illustrate some trend quite directly, without the need for a separate figure, as in: "over the last 60 years, the number of extreme weather events has continued to rise steadily, with some ups"  "and downs"  . Sparklines bring data graphics into the realm of typography and have the potential to make text, tables, social media, and smart devices more visually rich.

Ridgeline Plots

Playfair's graphic idea from the "Universal Commercial History" had still more to offer. Like many other good graphic ideas from history, it was recently rediscovered and given a new name, without recognition of Playfair as the source.

The difficulty with multiple time-series plots that share a common vertical axis is that they can crisscross in complicated ways, and it is hard to tell them apart. The simple device of using the same horizontal scale, but with the separate plots *offset vertically*, as in Plate 8, turned out to be a game changer for such data.

Plate 9 shows the yearly distributions of scores on a liberal (negative)—conservative (positive) dimension, based on all roll-call votes in the US House of Representatives and Senate from 1963 to 2013.[14] It is clear that the votes of US legislators have become increasingly polarized over time, but this shift has been asymmetric: the entire distribution of Republicans has shifted noticeably to the right, while that of Democrats has moved only slightly to the left.

Moreover, the two parties, particularly the Democrats, have become more homogeneous over time, in that the distributions of their scores have become narrower.

The remarkable feature of this graph is that so much can be seen in a relatively small space. It shows frequency distributions for two parties spanning fifty-one years, each one summarized by a smoothed curve (a density estimate). The vertical size of the graph is small, so there is a lot of overlap. A graphic trick, unavailable to Playfair, has been used to reduce this problem: the area underneath each curve was shaded with partially transparent colors, so that overlap can be seen as an increasing intensity of shading.

Why Playfair?

After considering the enormous explosion of graphical ideas and understandings that emerged fully realized from William Playfair's fertile mind and pen, it is natural to ask "Why Playfair?" and "Why then?" These entail the larger question, "Was it the man or the moment?" Were these developments inevitable at that time, so that if Playfair didn't come up with them, someone else would have? Or was it just Playfair's special vision and genius, that of a man far ahead of his time?

It is impossible to know for sure, but we believe that the answer almost surely is a combination of the man, the moment, and more. Problems recur, and solutions emerge when enough technology is available to allow a solution. And there must also be a person who cares about the problem and has the wit to use the technology. Last, the circumstances must be favorable enough for the solution to withstand attempts to suppress it.[15]

In answer to the question, "Why Playfair," clearly he had the idea of visualization from his brother's early instruction, and developed the skills to make graphs from his training as a draftsman. He learned to think visually about phenomena and events and had a remarkable eye for effective and visually pleasing designs.

Then there was the moment: the idea of learning from data was becoming prominent, though not yet accepted in graphic form. Playfair realized that he could use this medium to tell stories and make arguments far better than with mere words and numbers. The data fields of natural science and economics had been plowed and fertilized. They were now ready to be brought to market: for their popularization in a new form, designed to convey arguments and evidence directly to the eye.

This was evident when, in 1787, just two years before France was plunged into revolution, the Count of Vergennes delivered a gift to the royal court of Louis XVI, the last king of France before the fall of the monarchy. The package contained a copy of Playfair's *Commercial and Political Atlas*, published the previous year in London. Unlike conventional atlases, the volume contained no maps, but it did contain charts of a new and unfamiliar variety. Louis XVI, an amateur of geography and the owner of many fine atlases, examined his acquisition with great interest. Although the charts were novel, Louis had no difficulty in grasping their purpose. Many years later, Playfair wrote that the king "at once understood the charts and was highly pleased. He said they spoke all languages and were very clear and easily understood."[16]

Playfair's Legacy

Although we celebrate Playfair today as the father of modern data graphics, it also fair to say that Playfair's revolutionary graphical ideas were underappreciated in his lifetime. His elegant graphical innovations were often ignored and sometimes denigrated. His graphs of the national debt of England, for example, were criticized as "mere plays of the imagination."[17] Graphical images produced by recording devices of, say, barometric pressure (Figure 1.4), were reasonably well understood. However graphs of intangible data like the national debt seemed to stretch the imagination.[18] It was not until the last half of the nineteenth century that Playfair had much impact. In the United Kingdom, the noted economist William Stanley Jevons enthusiastically adopted Playfair's method.[19] Perhaps more importantly, Jevons influenced the great biometrician Karl Pearson to understand and lecture on the graphical method, as understood from Playfair.

In France, Playfair's influence could be seen in the work of Charles Joseph Minard (Plate 6) and Émile Cheysson, director for the *Album de Statistique Graphique* (Plate 7). In 1878 Étienne-Jules Marey, whom we will meet again in Chapter 9, extolled Playfair's methods in *La méthode graphique*, the first book on this topic.

By 1885, at the Silver Jubilee of the Statistical Society of London (now the Royal Statistical Society), graphical methods and visual reasoning to conclusions had become nearly mainstream. The noted political economist Alfred Marshall[20] addressed the attendees on the benefits of the graphic method in understanding economic trends, in words but not in pictures. Émile Levasseur[21] represented a French view, and presented a survey of the wide variety of graphs and statistical maps then in use. Playfair's ideas of a visual rhetoric and a means for explaining ideas graphically had finally taken hold.

The final, sad, and ignominious chapter of Playfair's life has only recently been uncovered and told by Ian Spence, Scott Klein, and Colin Fenn.[22] None of his grand and somewhat shady schemes paid off, and his books and pamphlets did not make him wealthy or perhaps even pay the rent. In his last few years he struggled particularly with debt and deteriorating health.

Playfair died on February 11, 1823, possibly from complications of advanced diabetes, and was interred in the Bayswater cemetery belonging to the church of St. George, Hanover Square. Spence and colleagues trace this and the subsequent history of this burial site, which is now occupied by the Lanesborough Hotel. Consequently, there is no marker for Playfair's tomb; but at least now historians of data visualization and fans of Playfair have a place to go in London and drop a token of appreciation, just as fans of Edith Piaf and Oscar Wilde do when they visit Père Lachaise cemetery in Paris.

6

The Origin and Development of the Scatterplot

As we saw in Chapter 5, most modern forms of data graphics—pie charts, line graphs, and bar charts—can generally be attributed to William Playfair in the period 1785–1805. All of these, even though presented as two-dimensional graphs, were essentially one-dimensional in their view of data. They showed a single quantitative variable (such as land area or value of trade) broken down by a categorical variable, as in a pie chart or bar chart, or plotted over time (perhaps with separate curves for imports and exports), as in a line graph.

In the development of a language and taxonomy of graphs, Playfair's graphs and other visual representations of data in this time can considered 1.5D—more than just a single variable shown, but not quite enough to qualify for 2D status. In Playfair's visual understanding, the horizontal axis in his plots most often bound to time, forcing him to use other means to show relations with other variables.

The next major invention in data graphics—the first fully two-dimensional one—was the *scatterplot*. Indeed, among all forms of statistical graphics, the scatterplot may be considered the most versatile and generally useful invention in the entire history of statistical graphics.[1]

Essential characteristics of a scatterplot are that two quantitative variables are measured on the same observational units (workers); the values are plotted as points referred to perpendicular axes; and the goal is to show something about the relation between these variables, typically how the ordinate variable, y, varies with the abscissa variable, x.

Figure 6.1 shows a typical, if simplistic, modern scatterplot. It relates the number of years of experience of some workers on the horizontal (x) axis to their current annual salary on the vertical (y) axis. The experience and salary

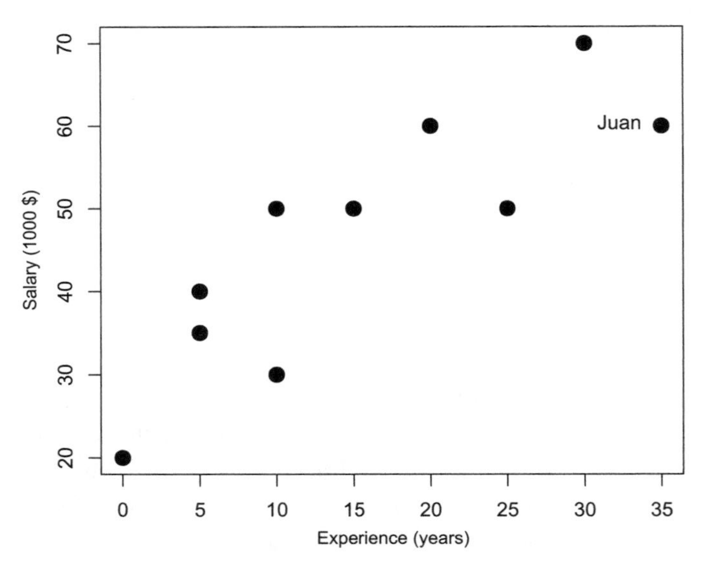

6.1 **Modern scatterplot:** A conventional modern scatterplot depicting the relation between salary and years of experience for a hypothetical group of ten workers. *Source:* © The Authors.

of each of ten workers are shown by a point at the (x, y) coordinates in this plot. One goal of such a plot would be to answer a question like, "How does salary depend on experience?" Modern statistical and graphic methods could provide some answers and also show something about the uncertainty attached to any such assessment.

A simple "reading" of Figure 6.1 is that salary increases with years of experience. But a virtue of the scatterplot is that it becomes a portal to deeper questions about relationships that can be answered graphically: (a) Is the relationship reasonably linear? Draw a line or a smooth curve on the plot. (b) Is there sufficient evidence in the data that increasing experience leads to greater salary? Draw some confidence bounds for a fitted relation. (c) If I work for fifty years, what can I expect my salary to be? Project the line or curve to fifty years. (d) Are men and women paid equally? Use different point symbols or colors to distinguish them. (e) Are there any observations to mention in my narrative? Label them.

An example such as Figure 6.1, the modern form of the scatterplot, took a long time to develop. Its form did not resemble the modern graph until the work of Francis Galton [1822–1911] on the heritability of traits. After this time, readers could perhaps begin to see, and reason from, pictures of dots inside a box such as that shown in Figure 6.1. The Cartesian reference frame, particularly for maps, had long been familiar. But it took something more for this to be viewed as a form of visual explanation for scientific phenomena. In this, Galton started with the problem of how to understand the inheritance of traits; he gained insight through statistical diagrams; and these became the source of the statistical ideas of correlation and regression and thus much of modern statistical methodology.

Early Displays That Were Not Scatterplots

The history of statistical graphics includes quite a few displays that resemble a scatterplot in some respects but don't quite meet our definition. The first prerequisite for a scatterplot was the idea of a *coordinate system*. Abstract, mathematical coordinate systems and the relations between graphs and functional equations, $y = f(x)$, such as the linear equation of a line, $y = a + bx$, were introduced in the 1630s (by Descartes and Fermat). The idea of two-dimensional, map-based coordinates, as systematized by Mercator, had been used since antiquity. What was new here was that Cartesian geometry introduced the idea of an abstract (x, y) plane, where equations could characterize all kinds of functional relations whose properties could be studied mathematically, in what is now called *analytic geometry*.

In the 1660s, the first proto–line graphs, showing weather data (barometric pressure) recorded over time, were introduced by Robert Plot (see Figure 1.4). Plot called this a "history of the weather," but it is little more than a tracing recorded by a pen on a moving paper chart. In 1669, Christiaan Huygens plotted survival versus age from John Graunt's data (see Figure 1.5), producing the first graph of an empirical continuous distribution function. This was an early example of turning a table into a graph, but not yet a scatterplot.

In 1686, Edmund Halley prepared the first known bivariate plot (perhaps derived from observational data, but not showing the data directly) of a theoretical curve relating barometric pressure to altitude (see Figure 6.2). The curve is simply a hyperbola, showing the inverse mathematical relation between these variables. The labeled horizontal and vertical lines attest to Halley's effort to explain visually how pressure decreases with altitude. Although

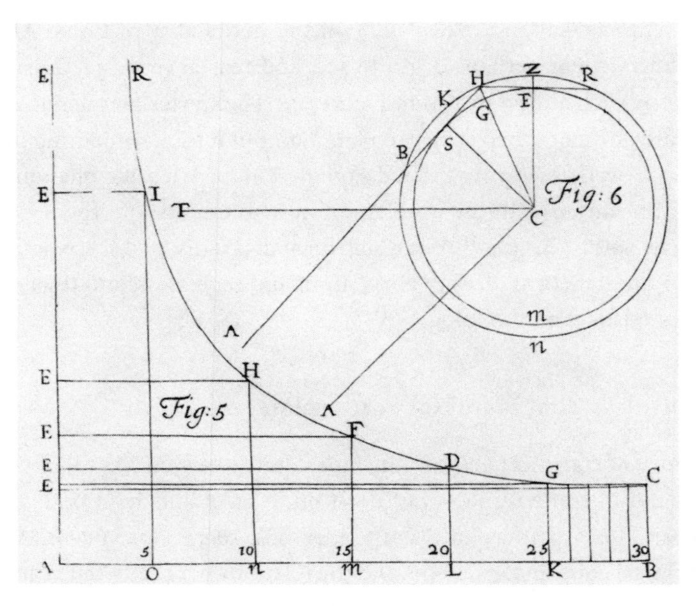

6.2 **A theoretical relation:** Edmund Halley's 1686 bivariate plot of the theoretical relation between barometric pressure (y) and altitude (x), derived from observational data. *Source:* Edmond Halley, "On the height of the mercury in the barometer at different elevations above the surface of the earth, and on the rising and falling of the mercury on the change of weather," *Philosophical Transactions*, 16, (1686), 104–115.

he shows numerical values for the altitudes on the horizontal axis, the vertical axis has none.

By the early 1700s many astronomers had begun to collect observational data on important scientific problems: What were the orbits of Jupiter and Saturn? What was the shape of the Earth: a perfect sphere? a flattened or elongated spheroid? The mathematical equations for these physical problems had been derived by the best mathematicians of the time, along with some ways to connect the data that could be observed on Earth into answers.

A key subproblem was how observations under possibly different conditions or by different observers should best be combined to give the most precise estimate. Astronomers often took some sort of average, but only when they considered the observations to be equally reliable. The problem of the "combination of observations" attracted the attention of Tobias Mayer, Roger Joseph Boscovitch, and Pierre-Simon Laplace in the mid-1700s, but mostly for

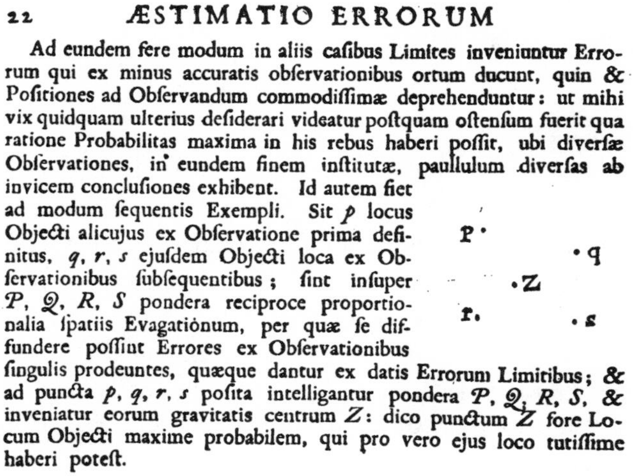

22 ÆSTIMATIO ERRORUM

Ad eundem fere modum in aliis caſibus Limites inveniuntur Errorum qui ex minus accuratis obſervationibus ortum ducunt, quin & Poſitiones ad Obſervandum commodiſſimæ deprehenduntur: ut mihi vix quidquam ulterius deſiderari videatur poſtquam oſtenſum fuerit qua ratione Probabilitas maxima in his rebus haberi poſſit, ubi diverſæ Obſervationes, in eundem finem inſtitutæ, paullulum diverſas ab invicem concluſiones exhibent. Id autem fiet ad modum ſequentis Exempli. Sit *p* locus Objecti alicujus ex Obſervatione prima definitus, *q*, *r*, *s* ejuſdem Objecti loca ex Obſervationibus ſubſequentibus; ſint inſuper *P*, *Q*, *R*, *S* pondera reciproce proportionalia ſpatiis Evagationum, per quæ ſe diffundere poſſint Errores ex Obſervationibus ſingulis prodeuntes, quæque dantur ex datis Errorum Limitibus; & ad puncta *p*, *q*, *r*, *s* poſita intelligantur pondera *P*, *Q*, *R*, *S*, & inveniatur eorum gravitatis centrum *Z*: dico punctum *Z* fore Locum Objecti maxime probabilem, qui pro vero ejus loco tutiſſime haberi poteſt.

p · q · Z · r · s

6.3 **Cotes's diagram:** Roger Cotes's depiction of four two-dimensional fallible observations which could be combined using a weighted mean at their center of gravity, Z, to give a more precise estimate. *Source:* Roger Cotes, *Aestimatio Errorum in Mixta Mathesis, per Variationes Planitum Trianguli Plani et Spherici*, 1722.

simple one-dimensional problems. This would later evolve into the method of least squares by Gauss and Euler around 1805.

Roger Cotes [1682–1716], a Cambridge mathematician who worked closely with Isaac Newton, had the idea to combine observations of different precision using a weighted mean, with weights inversely proportional to their possible errors.[2] Figure 6.3 shows a diagram he used in *Aestimatio Errorum*, which was published posthumously in 1722. He says in the text that if there are four observations, *p*, *q*, *r*, and *s*, and these are given weights *P*, *Q*, *R*, and *S*, then the best estimate to be taken from them would be their weighted mean, which is geometrically the center of gravity, the point shown as *Z*. This argument clearly required a two-dimensional diagram. Once again, this has features of a scatterplot, but Cotes clearly did not intend it as such.[3]

Toward the end of the eighteenth century, other developments led in the direction of the scatterplot: in 1794, a London physician, Dr. Buxton (about whom little is known), took out a patent and began to sell the first printed graph paper, with a rectangular coordinate grid. In 1796, John Southern and James Watt devised a pen-driven apparatus for the *automatic* recording of two variables simultaneously—pressure and volume in steam engines. Both ideas were substantial contributions to the idea of a scatterplot.

Johann Lambert

Between 1760 and 1777, Johann Heinrich Lambert [1728–1777] described curve fitting and interpolation from empirical data. Lambert, a Swiss polymath who made many contributions to mathematics, astronomy, color theory, and experimental sciences, was one of the first scientists to use graphs to represent experimental data, with an aim to show how algebraic methods could be applied to hand-drawn curves representing empirical observations.[4] He was a seeker of the mathematical laws governing physical phenomena.

Figure 6.4 shows a chart of soil temperatures in degrees Fahrenheit at a range of latitudes (individual curves) at some intervals of time over the year. The curves were derived from observational data, but no data points are shown. It is a fine example of an early graph, and it very clearly shows the phenomenon he sought to depict—very little variation at the equator and much greater variation toward the poles.

Yet a closer reading of Lambert's works shows that he had the essential ideas of the scatterplot and should be considered one of the founding fathers of data visualization, particularly for scientific phenomena. In a variety of works on topics of mortality, physics (color, light, hygrometry), and astronomy from 1760 to 1780, he consistently used graphs of data in an attempt to develop theory from fallible observations in such a way as to deal with a theory of errors. In several works he describes his use of graphs in a way that could be considered modern. One particularly clear statement appeared in 1765:

> We have in general two variable quantities x, y, which will be collated with one another by observation, so that we can determine for each value of x, which may be considered as an abscissa, the corresponding ordinate y. Were the experiments or observations completely accurate these ordinates would give a number of points through which a straight or curved line should be drawn. But as thus is not so, the line deviates to a greater or lesser extent from the observational points. It must therefore be drawn in such a way that it comes as near as possible to its true position and goes, as it were, through the middle of the given points.[5]

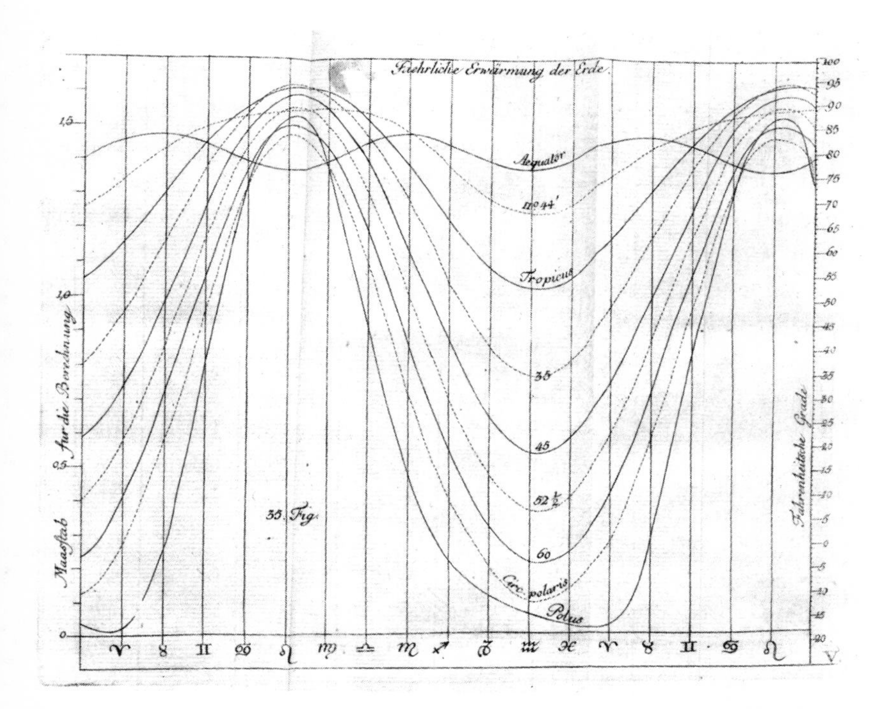

6.4 **Chart of soil temperatures:** J. H. Lambert's chart of the variation in soil temperatures at various latitudes over time. The horizontal axis represents time within a year using astronomical symbols. *Source:* Johann Heinrich Lambert, *Pyrometrie; oder, vom maasse des feuers und der wärme mit acht kupfertafeln.* Berlin: Haude & Spener, 1779.

The ideas expressed here were truly revolutionary. Not only does he describe the paradigm case of a plot of (x, y) points, but he also describes the idea of fitting a curve to represent "its true position . . . through the middle of the given points" as a way to account for errors of observation.

But if Lambert has not previously been credited as an originator of the scatterplot, it is perhaps ironically because the physical phenomena he was studying were so regular that, once the data points had been plotted, the smoothed curves alone could stand in as a representation of the points themselves, freed from error and avoiding visual clutter. Lambert's goals were even loftier. From Figure 6.4 and many others that he drew, his main desire was to characterize the lawlike regularity with mathematical equations.

Why Not Playfair?

Thus, well before 1800, all the necessary intellectual pieces for the graphing of empirical data on abstract 2D coordinate systems were in place. So, when Playfair devised nearly all of the common statistical graphs—first the line graph and bar chart in the *Commercial and Political Atlas*, later the pie chart and circle graph in the *Statistical Breviary*—one might wonder why he did not develop the scatterplot for data, the idea of plotting one variable against another.

Playfair was primarily concerned with economic data recorded over time, often for comparative purposes, so the time-series line graph seemed an ideal format. Indeed, all but one of the forty-four charts in the first edition of the *Commercial and Political Atlas* were line graphs, often showing two time series (imports and exports), so he could discuss the balance of trade as the difference between the curves.

The idea of plotting imports against exports apparently did not occur to him and probably would not have aided his arguments. In 1821 in a brief pamphlet, *Letter on Our Agricultural Distresses*, Playfair attempted something far more ambitious: to try to show the relationship between *different* time series and how these fit in terms of historical events. Plate 10 shows three parallel time series over a 250-year period, reflecting prices (price of a quarter of wheat in shillings), wages for labor (weekly wage for a good mechanic, in shillings), and the ruling British monarch.

Playfair shows the time series for wages as a line graph; its vertical scale on the left has a range of 0–100, but the data (wages) range only from 0 to 30. The time series for prices of wheat is shown as a bar graph using the right vertical scale, also with a range of 0–100.

Both vertical axes are in shillings, but the perception of their *relative* trends would change dramatically if the scales were changed from weekly to daily or monthly wages or if prices of wheat were changed to units of a loaf or a full bushel. But mixing different scales (y-axes), here wages and wheat prices, on the same plot is considered sinful today because it allows a sinning plotter to independently manipulate the two scales and make the relation between those two variables take any form they like.

Playfair's main goal was to show how spending power (wages) had changed in relation to buying power (prices) over two centuries and to lead directly to the perception that wheat (or bread) had become more affordable for workers. He concluded that: "the main fact deserving of consideration is, that never at any former period was wheat so cheap, in proportion to mechanical labor, as it is at the present time."[6]

But what the graph in Plate 10 actually shows *directly* is quite different. The strongest visual message is that wages changed relatively steadily (increasing very slowly up to the reign of Queen Anne and at a somewhat greater rate thereafter), whereas the price of wheat (and thus of bread and other items that could be purchased with those wages) fluctuated greatly. The inference that wages increased relative to prices toward the end is at best indirect and not visually compelling.

What Playfair wanted to do was to focus on the cost of wheat, relative to how much labor it took to buy it over time. To do this, he could have calculated a new, derived variable: the ratio of the price of wheat to wages, representing labor cost (number of weeks of work) required to buy a quarter-bushel of wheat. Plotting this derived variable over time, as in Figure 6.5, would have shown this directly, and illustrated his conclusion that wheat had become increasingly affordable for workers. In this modern version, we have added a smooth curve to show the overall trend. It highlights the fact that the ratio

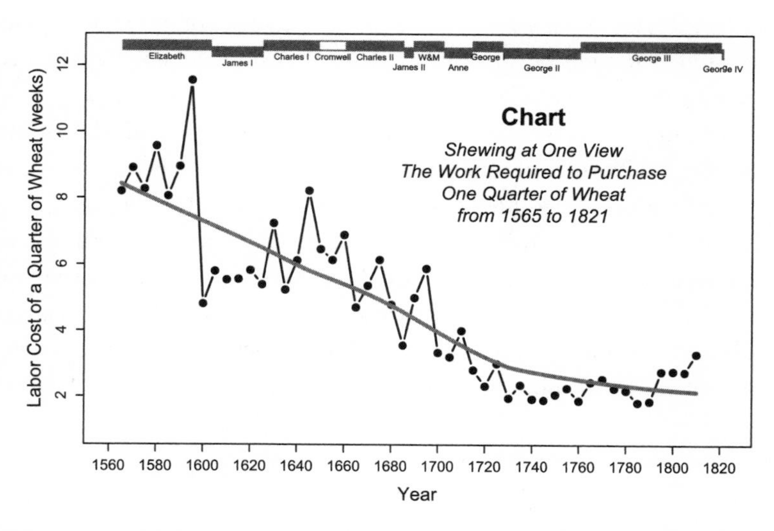

6.5 **Plotting the ratio:** Redrawn version of Playfair's time-series graph, showing the ratio of price of wheat to wages, together with a nonparametric smooth curve. *Source:* © The Authors.

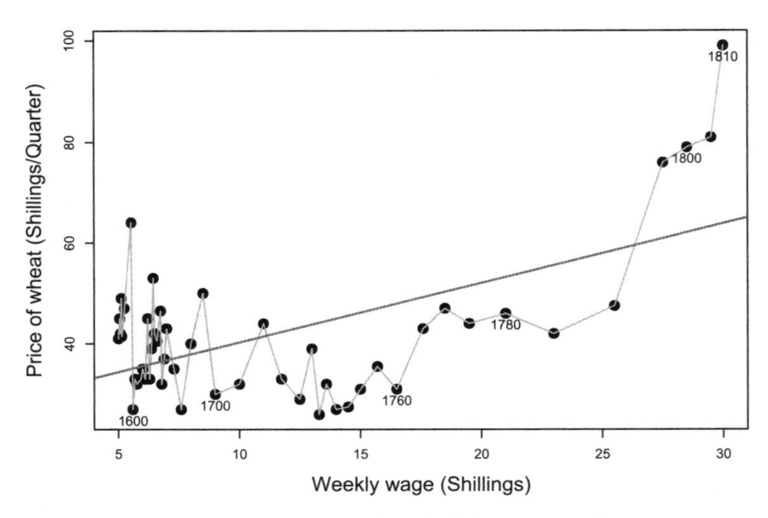

6.6 **Scatterplot view:** Scatterplot version of Playfair's data, joining the points in time order, and showing the linear regression of price of wheat on wages. *Source:* © The Authors.

of price to wages declined on average fairly steadily, leveling off in the most recent years "QED!".

It is reasonable to infer that Playfair, in nearly all his works, thought of his data as a collection of *separate* time series, each a series of numbers ordered by a dominant dimension of time. And there is no evidence that he ever considered plotting derived values, such as the ratio of price to wages.[7]

Similarly, in his charts of imports and exports (see Figure 5.5), he did not think to plot the "balance of trade," the difference of exports − imports, when this was *exactly* what he wanted to convey in his comparative charts.

If Playfair had used scatterplots for such data—for example, plotting price of wheat against wages—the result would not have supported his argument very well. Figure 6.6 shows a plot of price versus wages, with the points joined in time order. The linear regression line shows an increasing trend, reflecting the fact that both wages and the price of wheat rose over time.

John Herschel and the Orbits of Twin Stars

In the hundred years from 1750 to 1850, during which most of the modern graphic forms were invented, fundamentally important problems of measurement attracted the best mathematical minds, including Euler, Laplace,

Legendre, Newton, and Gauss, and led to the inventions of calculus, least squares, curve fitting, and interpolation.[8] In these scientific and mathematical domains, graphs had begun to play an increasing role in the explanation of scientific phenomena, as we described earlier in the case of Johann Lambert.

Among this work, we find the remarkable paper of Sir John Frederick W. Herschel [1792–1871], *On the Investigation of the Orbits of Revolving Double Stars*, which he read to the Royal Astronomical Society on January 13, 1832, and published the next year. Double stars had long played a particularly important role in astrophysics because they provided the best means to measure stellar masses and sizes, and this paper was prepared as an addendum to another 1833 paper, in which Herschel had meticulously cataloged observations on the orbits of 364 double stars.

The printed paper refers to four figures, presented at the meeting. Alas, the version printed in the *Memoirs of the Royal Astronomical Society* did not include them, presumably owing to the cost of engraving. Herschel noted, "The original charts and figures which accompanied the paper being all deposited with the Society."[9] These might have been lost to historians, but Thomas Hankins discovered copies of them in research for his insightful 2006 paper on Herschel's graphical method.

To see why Herschel's paper is remarkable, we must follow his exposition of the goals, the construction of scatterplots, and the idea of visual smoothing to produce a more satisfactory solution for parameters of the orbits of double stars than were available by analytic methods.

The printed paper began with a strong statement of his goals and achievement:

> My object, in the following pages, is to explain a process by which the elliptic elements of the orbits of binary stars may be obtained, from such imperfect observations as we actually possess of them, not only with greater facility, but also with a higher probability in their favour, than can be accomplished by any system of computation hitherto stated.[10]

Herschel noted that the laws of gravitation imply that the elliptic orbits of binary stars may be determined from the measured angles between the meridian and a line to their centers, and their apparent distances from each other, recorded over relatively long periods of time. Were these measurements exact, or even determined with relatively small errors, the well-known principles of elliptic motion and spherical trigonometry would provide precise solutions for the constants (seven in number) that specify the orbit and its relation over time to the position of an earthly observer. But, he noted that the angles and distances between double stars are measured with "extravagant errors," particularly in the distances, and previous analytic methods, which depend on solving seven equations in seven unknowns, had been unsatisfactory. He proclaimed his use of a better, graphical solution:

> The process by which I propose to accomplish this is one essentially graphical; by which term I understand not a mere substitution of geometrical construction and measurement for numerical calculation, but one which has for its object to perform that which no system of calculation can possibly do, by bringing in the aid of the eye and hand to guide the judgment, in a case where judgment only, and not calculation, can be of any avail. (Herschel 1833b, p. 178)

Herschel then described the process of constructing a sheet of graph paper "covered with two sets of equidistant lines, crossing each other at right angles, and having every tenth line darker than the rest." Then, points consisting of the angles of position (y) and date of observation (x) are plotted. "Our next step, then, must be to draw, by the mere judgment of the eye, and with a free but careful hand, not *through*, but *among* them, a curve presenting as few and slight departures from them as possible, consistently with this character of large and graceful sinuosity, which must be preserved at all hazards" (emphasis in the original).[11]

Herschel's use of the scatterplot and the role that visual smoothing played in his analysis can best be illustrated with his first example,[12] on the orbits of Gamma Virginis, the third-brightest (double) star in the constellation Virgo. Here he refers to the raw data[13] comprising eighteen observations of the position angle and separation distance for this double star over the period 1718–1830. His graph of these observations, shown in Figure 6.7, is our nominee for the first true scatterplot.

The (apparent) orbits of a twin star can be described completely by the position angle between the central, brighter star and its twin satellite star measured from the North Celestial Pole, and the angular separation distance between the two stars, measured in seconds of arc. The physical setting is

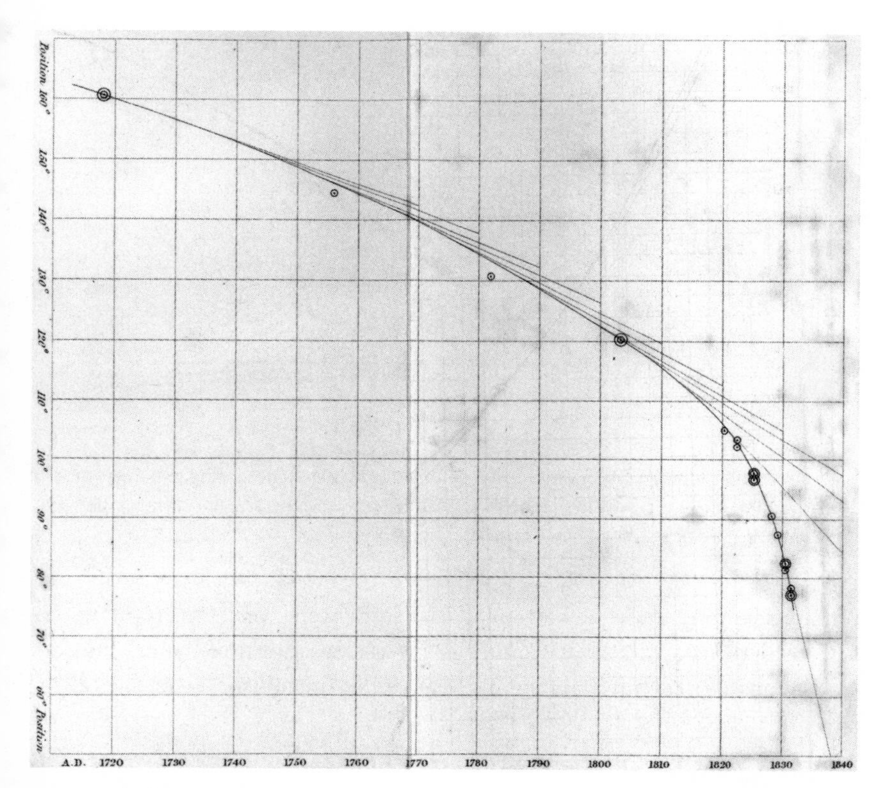

6.7 **Herschel's first scatterplot:** Herschel's graphical method applied to his data on the double star γ *Virginis*. The plot shows observations of position angle separating the twin stars on the vertical axis and time on the horizontal axis. Some observations, considered as more reliable, are shown as double circles. The key feature is the smoothed curve, from which he drew the tangent lines, giving him the angular velocities at evenly spaced distances along the curve. *Source:* John F. W. Herschel, "On the investigation of the orbits of revolving double stars: Being a supplement to a paper entitled 'micrometrical measures of 364 double stars,'" *Memoirs of the Royal Astronomical Society*, 5, (1833), 171–222.

sketched in Figure 6.8. One astronomical observation gave one point on the orbit. The relative positions of twin stars changed slowly, so over a long period of time, such measurements would give other points, which could be used to calculate the apparent orbit.

Herschel's problem was that the recorded data were incomplete and of varying accuracy: fourteen observations had position angle recorded and nine had measures of separation distance, but only five had both position angle and

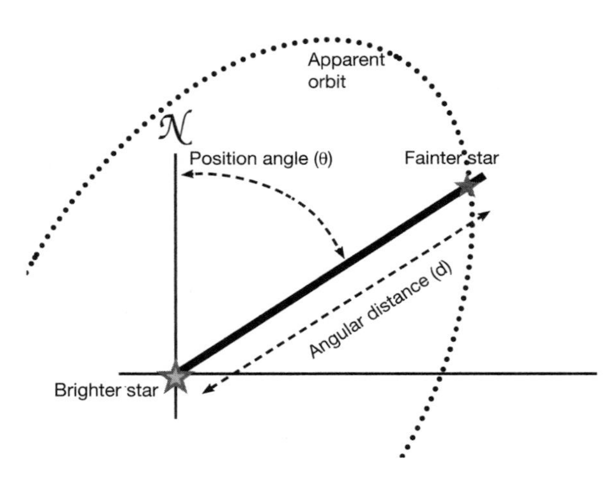

6.8 **Star measurements:** Observations of a double star involved measuring two quantities to determine the orbit: the position angle of the fainter star and the angular distance between them. *Source:* © The Authors.

distance. Among these, he had noted that some were "very uncertain; not to be considered as an observation" or "one night's measure; no reliance," indicating the possibly "extravagant errors" referred to previously (Herschel 1833a, table, p. 35).

To appreciate the role that even tiny errors might play, Herschel wrote later, in his book *Outlines of Astronomy* (1860), that a mere half-second error in the position angle meant that on a circle 6 feet in diameter, such an error would correspond to only 1/12,000 of an inch, far smaller than anything that could be measured with physical instruments.

Herschel's solution to these problems of both data and technique must be regarded as conceptually ingenious, and almost certainly the first case in which a scatterplot led to an answer to a scientific question. Rather than work with the position-distance pairs, for which only 5 points were available, he chose to work with the graph of position angle over time. Figure 6.9 shows a reconstruction[14] of the graph of his fourteen observations, annotated with the authority cited for each ("H" refers to Herschel's father, William, who discovered Uranus; "h" refers to Herschel himself).

However, he noted:

> But since equal reliance can probably not be placed on all the observations, we must take care to distinguish those points which correspond

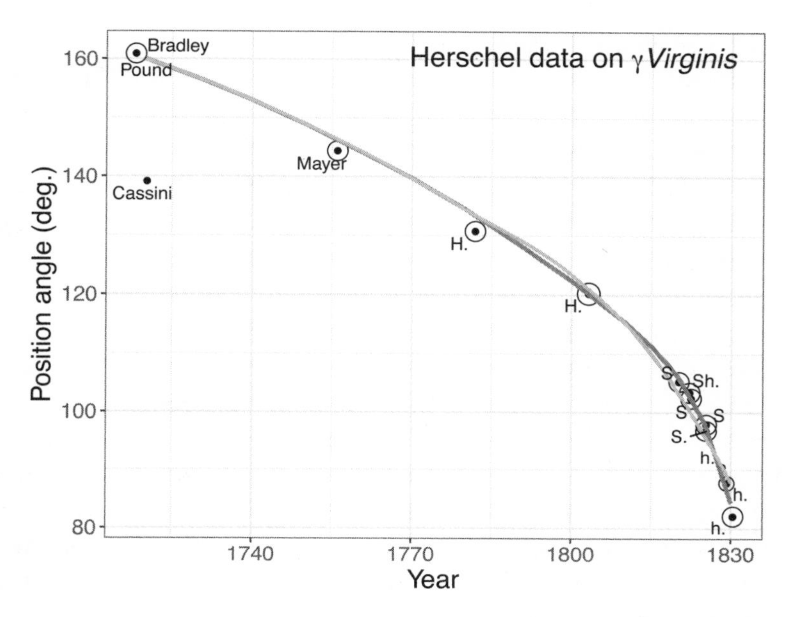

6.9 **Herschel redone:** Reconstruction of Herschel's graph of data on the orbits of γ *Virginis* together with his eye-smoothed, interpolated curve (light gray) and a loess smoothed curve (dark gray). Circles around each data point are of area proportional to Herschel's weight for the observation. *Source:* © The Authors.

> to observations entitled to the greatest confidence. . . . These should be marked on the chart in some special manner . . . so as to strike the eye at a general glance; for example, by larger or darker points. . . . And when we draw our curve, we must take care to make it pass either through or very near all those points which are thus most distinguished. (Herschel 1833b, p. 179)

Accordingly, we assigned a weight, in the range of 0.5 to 6, to each observation, reflecting our judgment of Herschel's confidence from his notes. The solid curve in Figure 6.9 shows Herschel's interpolated points (hollow circles), connected by line segments; it is apparent that he totally ignored the observation by Cassini (we gave it a weight of 0.5). To see how well we have captured Herschel's method, we also fit a statistically smoothed (loess) curve, using our weights. Over the latter range of years, where the observations are most dense, the two curves agree closely, but they differ markedly for the years 1730–1790. However, Herschel's curve, fit by "eye and hand to guide the judgment," is somewhat smoother ("of large and graceful sinuosity") than that found by modern non-parametric regression smoothing. He concluded, "This curve once drawn, must represent, it is evident, the law of variation of the angle of position, with the time, not only for the instants intermediate between the dates of the observations, but even *at the moments of observations themselves*, much better than the individual *raw* observations can possibly (on an average) do" (Herschel 1833b, p. 179; emphasis in the original).

His next step was to measure, from the graph, the angular velocity, $d\theta/dt$, by calculating the slopes of the curve at the interpolated points. These calculations are shown as the tangent lines to the curve in Figure 6.7. From these, he could calculate measures of separation distance, "independent altogether of direct measurement," as distance $\sim 1/\sqrt{d\theta/dt}$, because in either the real or apparent ellipse of motion, the areas swept out over time must be proportional to time, so the distances are inversely proportional to the square roots of angular velocities. Finally, he could plot the smoothed ellipse of the apparent orbit, and thus calculate the parameters that determine the complete motion of Gamma Virginis.

The result of all this work is shown in Figure 6.10. In this figure Herschel carefully shows the interpolated data points from 1720 to 1830 derived from Figure 6.7 and translated to this space. The proof of his method was largely visual: the calculations from his smoothed curve gave a nearly perfect ellipse.[15] One can appreciate the patience and precision required, because the complete orbit would take about 600 years to fully observe and Herschel had only about 100 years of data. We can imagine Herschel finishing his lecture by writing "QED!" on the chalkboard and the audience rising with thunderous applause.

Thus, a problem that had confounded astronomers and mathematicians for at least a century yielded gracefully to a graphical solution based on a scatterplot. The crucial steps were smoothing the raw observations and understanding that theory could go far beyond the available data.

Herschel's Graphical Impact Factor

The critical reader may object, thinking that Herschel's graphical method, as ingenious as it might be, did not produce true scatterplots in the modern sense because the horizontal axis in Figure 6.9 is time rather than a separate variable. Thus one might argue that all we have is another time-series graph,

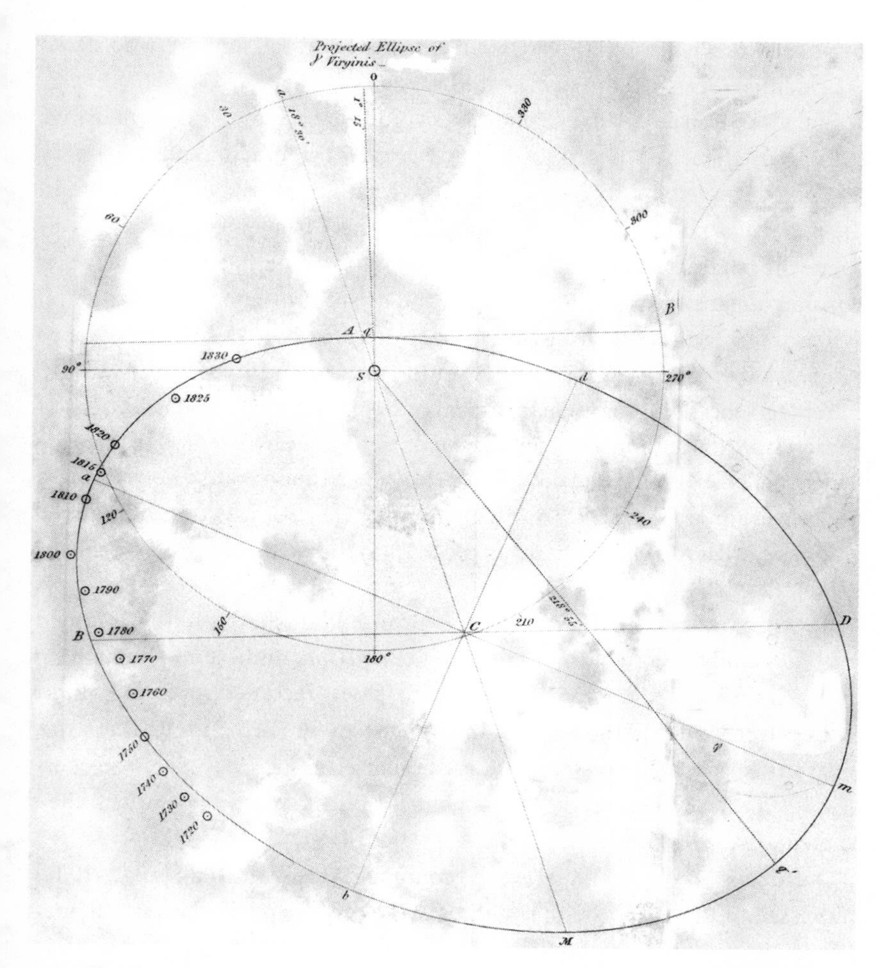

6.10 **Ellipse geometry:** Herschel's geometric construction of the apparent elliptical orbit of Gamma Virginis from the calculations based on Figure 6.7. The principal star is marked "S." The labeled points around the ellipse show the interpolated observations translated to this space. *Source:* John F. W. Herschel, "On the investigation of the orbits of revolving double stars: Being a supplement to a paper entitled 'micrometrical measures of 364 double stars,'" *Memoirs of the Royal Astronomical Society*, 5, (1833), 171–222.

so priority really belongs to Playfair, or further back, to Lambert, who stated the essential ideas. On the surface this is true.

But it's only true on the surface. We argue that a close and appreciative reading of Herschel's description of his graphical method can, at the very least, be considered a true innovation in visual thinking, worthy of note in the present account. More importantly, Herschel's true objective was to calculate the parameters of the orbits of twin stars based on the relation between position angle and separation distance; the use of time appears in the graph as a proxy or indirect means to overcome the scant observations and perhaps extravagant errors in the data on separation distance.

Yet Herschel's graphical development of calculation based on a scatterplot attracted little attention outside the field of astronomy, where his results were widely hailed as groundbreaking in the Royal Astronomical Society. But this notice was for his scientific achievement rather than for his contribution of a graphical method, which scientists probably rightly considered just a means to an end.

It took another 30–50 years for graphical methods to be fully welcomed into the family of data-based scientific explanation, and seen as something more than mere illustrations. This change is best recorded in presentations at the Silver Jubilee of the Royal Statistical Society in 1885. Even at that time, most British statisticians still considered themselves "statists," mere recorders of statistical facts in numerical tables; but "graphists" had finally been invited to the party.

On June 23, the influential British economist Alfred Marshall [1842–1924] addressed the attendees on the benefits of the graphic method, a radical departure for a statist. His French counterpart Émile Levasseur [1828–1911] presented a survey of the wide variety of graphs and statistical maps then in use. Yet even then, the scientific work of Lambert and Herschel, and the concept of the scatterplot as a new graphical form remained largely unknown. This would soon change with Francis Galton.

Francis Galton and the Idea of Correlation

Francis Galton [1822–1911] was among the first to show a purely empirical bivariate relation in graphical form using actual data with his work on questions of heritability of traits. He began with plots showing the relationship

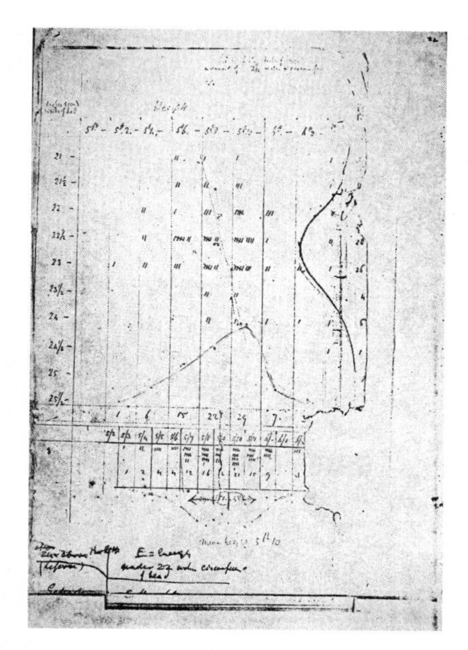

6.11 **The first correlation diagram:** Galton's first correlation diagram, showing the relation between head circumference and height, from his undated notebook, "Special Peculiarities." *Source:* Reproduced from Victor L. Hilts, *A Guide to Francis Galton's English Men of Science*, Vol. 65, Part 5. Philadelphia, PA: Transactions of the American Philosophical Society, 1975, pp. 1–85, fig. 5.

between physical characteristics of people (head circumference and height) or between parents and their offspring, as a means to study the association and inheritance of traits: Do tall people have larger heads than average? Do tall parents have taller than average children?

Inspecting and calculating from his graphs, he discovered a phenomenon he called "regression toward the mean," and his work on these problems can be considered to be the foundation of modern statistical methods. His insight from these diagrams led to much more: the ideas of correlation and linear regression; the bivariate normal distribution; and eventually to nearly all of classical statistical linear models (analysis of variance, multiple regression, etc.).

The earliest known example is a chart of head circumference compared to stature from Galton's notebook (circa 1874) "Special Peculiarities," shown in Figure 6.11.[16] In this hand-drawn chart, the intervals of height are shown horizontally against head circumference vertically. The entries in the body are the tallies in the pairs of class intervals. Galton included the total counts of height and head circumference in the right and bottom margins and drew smooth curves to represent their frequency distributions. The conceptual origin of this chart as a table rather than a graph can be seen in the fact that the smallest values of the two variables are shown at the top left (first row and first column), rather than in the bottom right, as would be more natural in a graph.

One may argue that Galton's graphic representations of bivariate relations were both *less* and *more* than true scatterplots of data, as these are used today. They are less because at first glance they look like little more than tables with some graphic annotations. They are more because he used these as devices to calculate and reason with.[17] He did this because the line of regression he sought was initially defined as the trace of the mean of the vertical variable y as the horizontal variable x varied[18] (what we now think of as the conditional mean function, $\mathcal{E}(y \mid x)$), and so required grouping at least the x variable into class intervals. Galton's displays of these data were essentially graphic transcriptions of these tables, using count-symbols (/, //, ///, . . .) or numbers to represent the frequency in each cell—what in 1972 the Princeton polymath John Tukey called "semi-graphic displays," making them a visual chimera: part table, part plot.

Even though this chimera counts only as a quick and dirty scatterplot, the step from a bivariate frequency table to a semigraphic display was a crucial one for Galton, and for the history of statistics and statistical graphics as well. As Tukey pointed out, one key difference between a semi-graphic display (the tallies in the bottom of Figure 6.11) and the fully graphic form (frequency polygon) is that one can compute reasonably well from the former but not from the latter. The faint diagonal lines in the figure show Galton's initial sketch of how to calculate the average trend of head circumference from height.

Galton's next step followed a method he had developed in earlier work (discovering weather patterns in 1861; see the section "Francis Galton's Greatest Graphical Discovery" in Chapter 7) of crowd-sourced data collection; it was also one of the earliest examples of the design of an experiment to control extraneous factors.

In 1875, Galton distributed packets of sweet pea seeds to each of seven friends, with instructions to carefully grow them all, harvest the seeds from

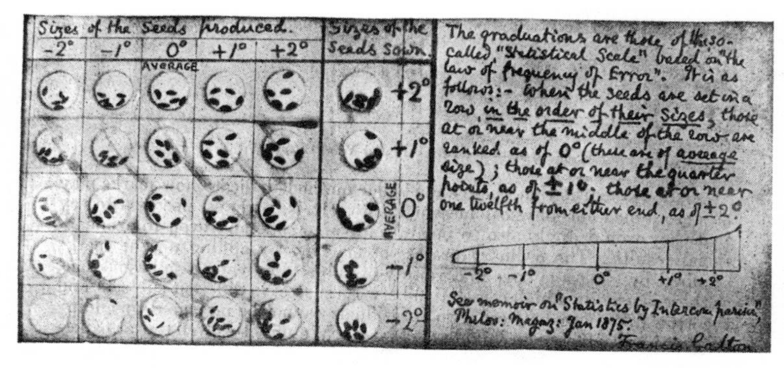

6.12 **Sweet pea diagram:** Galton's semigraphic table of the sweet pea data, represented in classed intervals around the averages for parent and child seeds. The marks in the cells are supposed to represent the number of seeds in each combination. *Source:* galton.org.

the new generation and return the child seeds to him.[19] Each friend received seven packets of seeds, each packet containing 10 seeds of approximately the same size / weight. The seven packets varied roughly uniformly in size, and were identified only by letters, "K" through "Q." Thus, he sent out $7 \times 7 \times 10 =$ 490 seeds in what would be called today a two-way factorial experiment, with 7 friends, and 7 treatments (sizes) as factors. When his friends returned the harvested child seeds (separately for each packet) he carefully measured the sizes of the seeds in returned packets and tabulated them against the sizes of the parent seeds.[20]

The semigraphic display shown in Figure 6.12 is his first attempt to visualize this relationship. As the side note explains, he was principally interested in showing that the distributions of sizes of both the seeds sown and their offspring followed the "law of frequency of error" (the normal distribution).

In 1877, for a lecture at the Royal Institution, Galton made a proper graph of the average size of offspring seeds produced from parent seeds in the various size classes. The original may be lost, but Karl Pearson, in his multivolume *The Life, Letters and Labours of Francis Galton*, reproduced this graph (Volume 3A, Chapter 14, fig. 1) and titled it "the first regression line." Figure 6.13 is a modern reconstruction of that figure, which also shows the actual data on which it was based[21] and illustrates Galton's reasoning.

Galton saw that the average size of child seeds from a particular size of parent seed could be approximately described by a straight line. This was one

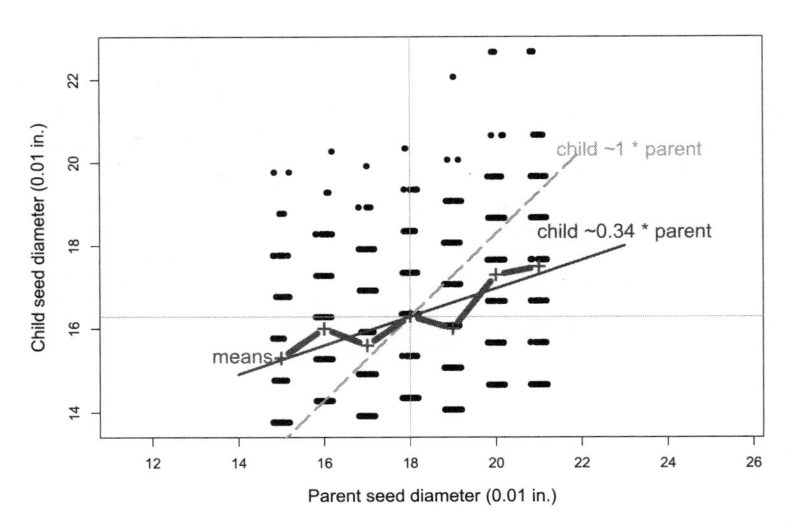

6.13 **Visual argument for regression:** Reconstruction of Galton's argument from the experiment on sweet peas, from his 1877 graph. The plotted points show the data that Galton later reported in his 1889 *Natural Inheritance* (table 2, p. 226), but jittered horizontally because the sizes of the parent seeds were discrete. The solid line shows the fitted linear relation of means (+), joined by a thicker line. The fact that the fitted line of the means had a slope less than 1.0 implied what Galton later called "regression toward the mean" in filial characteristics. *Source:* © The Authors.

key insight. If he had stopped there, this would have been enough to be noteworthy in the history of data visualization as a way to describe a nearly linear relation between two observed variables. But Galton's insight was far more powerful, because he was concerned with the issue of heredity of traits. If heredity was all-important, one would expect that the average size of offspring would deviate from the mean *exactly* as did the parent seeds. That is, the slope of the line of means would be 1.0, as shown by the gray line in Figure 6.13.

But the slope of the line of means was far less than 1, approximately 1/3, and Galton was able to conclude something far more interesting and general: a "reversion toward mediocracy," later called "regression toward the mean." He said:

> offspring did not tend to resemble their parent seeds in size, but to be always more mediocre than they—to be smaller than the parents, if the

> parents were large; to be larger than the parents, if the parents were very small. (Galton 1886, p. 246)

In 1900 Karl Pearson worked out the statistical theory behind such linear relations and invented a measure of the strength of the correlation that we now know as "Pearson's *r*." Pearson acknowledged that this symbol *r* was first used by Galton to indicate a proportional "coefficient of reversion" in a partially genetic process. In Figure 6.13, the slope of the line of means is 0.34; in this case that value is also the correlation between parent and child seed sizes. In modern parlance, one could say that the parent size accounts for $r^2 = (0.34)^2 = 0.12$, or 12 percent of the variation in the size of child seeds.

Galton's Elliptical Insight

Galton's next step on the problem of filial correlation and regression turned out to be one of the most important in the history of statistics. In 1886, he published a paper titled "Regression Towards Mediocrity in Hereditary Stature" containing the table shown in Figure 6.14. The table records the frequencies of the heights of 928 adult children born to 205 pairs of fathers and mothers, classified by the average height of their father and mother ("mid-parent" height).[22]

If you look at this table, you may see only a table of numbers with larger values in the middle and some dashes (meaning 0) in the upper left, and bottom right corners. But for Galton, it was something he could *compute* with, both in his mind's eye and on paper.

> I found it hard at first to catch the full significance of the entries in the table, which had curious relations that were very interesting to investigate. They came out distinctly when I "smoothed" the entries by writing at each intersection of a horizontal column with a vertical one, the sum of the entries in the four adjacent squares, and using these to work upon. (Galton, 1886, p. 254)

Consequently, Galton first smoothed the numbers in this table, which he did by the simple step of summing (or averaging) each set of four adjacent

	Height of adult child												
Height of mid-parent	<61.7	62.2	63.2	64.2	65.2	66.2	67.2	68.2	69.2	70.2	71.2	72.2	73.2
>73.0	–	–	–	–	–	–	–	–	–	–	–	1	3
72.5	–	–	–	–	–	–	–	1	2	1	2	7	2
71.5	–	–	–	–	1	3	4	3	5	10	4	9	2
70.5	1	–	1	–	1	1	3	12	18	14	7	4	3
69.5	–	–	1	16	4	17	27	20	33	25	20	11	4
68.5	1	–	7	11	16	25	31	34	48	21	18	4	3
67.5	–	3	5	14	15	36	38	28	38	19	11	4	–
66.5	–	3	3	5	2	17	17	14	13	4	–	–	–
65.5	1	–	9	5	7	11	11	7	7	5	2	1	–
64.5	1	1	4	4	1	5	5	–	2	–	–	–	–
<64.0	1	–	2	4	1	2	2	1	1	–	–	–	–
Totals	5	7	32	59	48	117	138	120	167	99	64	41	17
Medians	–	–	66.3	67.8	67.9	67.7	67.9	68.3	68.5	69.0	69.0	70.0	–

6.14 **Semigraphic table:** Galton's Table I on the relationship between heights of parents and their children. Entries in the table are the number of adult children. *Source:* Reformatted from Francis Galton, "Regression towards Mediocrity in Hereditary Stature," *Journal of the Anthropological Institute of Great Britain and Ireland*, 15 (1886), 246–263, table I.

cells. We can imagine that he wrote that average number larger in red ink, exactly at the intersection of these four cells. When he had completed this task, we can imagine him standing above the table with a different pen and trying to connect the dots—to draw curves, joining the points of approximately equal frequency. We tried to reproduce these steps in Figure 6.15, except that we did the last step mechanically, using a computer algorithm, whereas Galton probably did it by eye and brain, in the manner of Herschel, with the aim that the resulting curves should be gracefully smooth.

He could clearly see that the frequencies increased toward the middle (the average heights of parents and their children). But more importantly,

> I then noticed that lines drawn through the entries of the same value formed a series of similar and concentric ellipses.... The points where

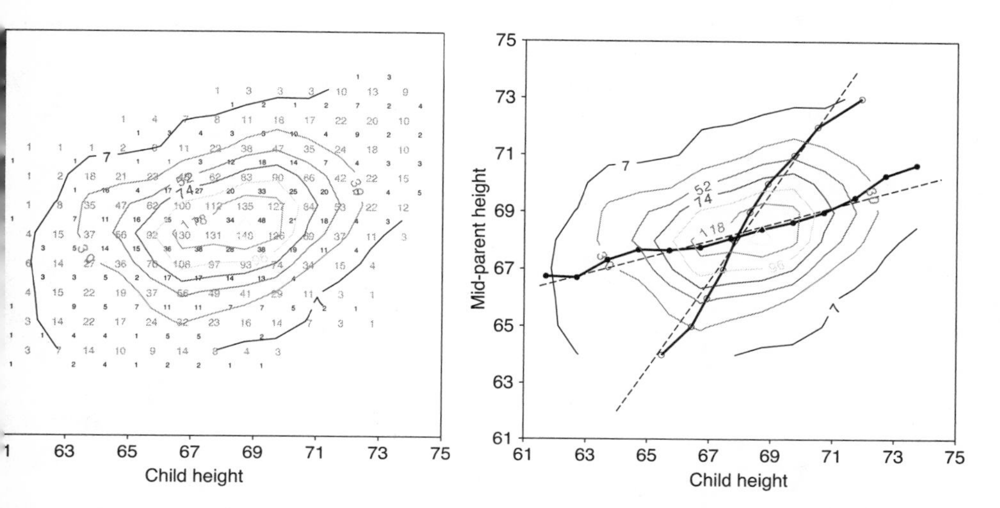

6.15 **Finding contours:** A reconstruction of Galton's method for finding contours of approximately equal frequency in the relation between heights of parents and their children. Left: The original data from his table (Figure 6.14) are shown in a small font (dark); the smoothed values are shown in a large font (light). Right: the lines joining the means of $y|x$ and the means of $x|y$ are added, together with the corresponding regression lines. *Source:* © The Authors.

> each ellipse in succession was touched by a horizontal tangent, lay in a straight line inclined to the vertical in the ratio of 2/3; those where they were touched by a vertical tangent lay in a straight line inclined to the horizontal in the ratio of 1/3. (pp. 254–255)

Galton illustrated the result of this visual insight in the diagram shown in Figure 6.16. These "similar and concentric ellipses" are now called contours of the bivariate normal distribution, the 3D frequency diagram of two correlated normally distributed variables. They are also called data (or concentration) ellipses, and they are used today as a visual summary of a scatterplot. They have remarkable properties for understanding statistical relations.[23]

The right panel of Figure 6.15 shows a reconstruction of the work leading to another aspect of Galton's visual insight from these similar and concentric ellipses. He calculated and plotted the means of $y \mid x$ and the means of $x \mid y$ and noted that they were tolerably close to the lines of horizontal

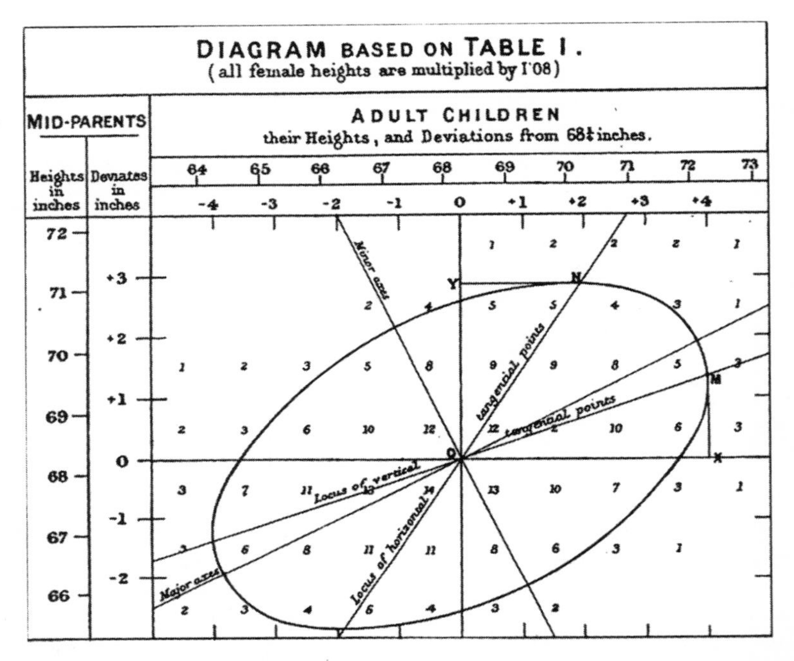

6.16 **Elliptical insight:** Galton's smoothed correlation diagram for the data on heights of parents and children, showing one ellipse of constant frequency. The geometric relations between the regression lines and the tangent lines and major and minor axes are also shown. *Source:* Extracted from Francis Galton, "Regression towards Mediocrity in Hereditary Stature," *Journal of the Anthropological Institute of Great Britain and Ireland*, 15 (1886), 246–263, Plate X.

and vertical tangency to the ellipses. Figure 6.16 shows the geometric relations Galton discovered from his graphical analysis of heredity of height. The loci of vertical and horizontal tangent lines (called conjugate diameters of an ellipse) turned out to have clear statistical interpretations as the regression lines, or lines for predicting y from x and of x from y, respectively. The major and minor axes of the ellipses, shown in Galton's figure, correspond to the principal components of the data, a relation discovered in 1901 by Karl Pearson.[24]

These geometric relationships would soon give rise to a number of important discoveries and inventions. As early as 1877, Galton had used the term "reversion," and then later "regression," to refer to the tendency of offspring to have a nearly linear relation to the same characteristic of their parents, with a slope, $r < 1$. He found it curious and also intriguing that the slopes of the lines formed by the points of tangency were in inverse relation, 1/3 : 2/3, and also that the slope for y on x was nearly the same as the value he had found for the sweet pea data.

Then, between the late fall of 1888 and 1890, Galton had another epiphany, somewhat akin to the story of Newton and the apple whose descent launched the theory of gravitation. In his earlier work with sweet peas, he had linked the phenomenon of slope of means < 1 to a hereditary component he called "reversion toward mediocracy," as illustrated in Figure 6.13.

From consideration of his Table I, shown graphically in Figure 6.16, Galton could see that the heights of children and those of mid-parents are treated symmetrically and that the slopes of *both* lines formed by the points of tangency were less than one. Thus, there is "reversion" in each direction: the reversion toward the mean height of children with respect to their mid-parents, and also the reversion on the side of mid-parents with respect to their children. It seemed clear that the phenomenon of reversion has nothing to do with heredity—the concept of reversion can only work forward in time! But what, if not heredity?

There was still more fruit on this tree: the concept of correlation.[25] In late 1888, Galton was also working on two seemingly unrelated problems: How can an anthropologist estimate the total height of an individual from a single bone (e.g., a thigh bone) recovered from a grave? How can criminal investigators use partial measurements (e.g., the size of a footprint) to identify the size or weight of an individual?

In a flash, he realized that these were just other instances of the problem he had observed in the heights of parents and their children and in the sizes of peas and their progeny. In an article published in 1890, "Kinship and Correlation,"[26] he announced the solution to this problem: the phenomenon of regression toward the mean is largely a statistical one: whenever two quantitative variables are imperfectly related ($|r| < 1$), regression toward the mean will be observed.

> Reflection soon made it clear to me that not only were the two new problems identical in principle with the old one of kinship which I had already solved, but that all three of them were no more than special cases of a much more general problem—namely, that of Correlation.

The mathematical theory of correlation and its relation to the bivariate normal distribution would shortly be worked out by others, including Pearson. In his 1920 paper on the history of correlation, Pearson gave a fitting appreciation of Galton's contribution:

> that Galton should have evolved all this . . . is to my mind one of the most note-worthy scientific discoveries arising from analysis of pure observation. (Pearson, 1920, p. 37)

Another Asymmetry

There is still one more small, but nagging, problem with this description of Galton's development of regression and the idea of correlation. In Figure 6.13, which shows Galton's sweet pea data, we were careful to plot the size of child seeds on the vertical y axis against that of their parent seeds on the horizontal x axis, as is the modern custom for a scatterplot, whose goal is to show how y depends on, or varies with, x. Modern statistical methods that flow from Galton and Pearson are all about *directional* relationships, and they try to predict y from x, not vice-versa. It makes sense to ask how a child's height is related to that of its parents, but it stretches the imagination to go in the reverse direction and contemplate how a child's height might influence that of its parents.

So, why didn't Galton put child height on the y axis and parent height on the x axis in Figure 6.16, as one would do today? One suggestion is that such graphs were in their infancy, so the convention of plotting the outcome variable on the ordinate had not yet been established. Yet in Playfair's time-series graphs (Plate 10) and in all other not-quite-scatterplots such as Halley's (Figure 6.2), the outcome variable was always shown on the y axis.

The answer is surely that Galton's Figure 6.16 started out as a table, listing mid-parent heights in the rows and heights of children in the columns. Parent height was the first grouping variable, and he tallied the heights of their children in the columns.

In a table, the rows are typically displayed in increasing order (of y) from top to bottom; a plot does the reverse, showing increasing values of y from bottom to top. Hence, it seems clear that Galton constructed his Table I (Figure 6.14) and figures based on it (Figure 6.15 and Figure 6.16) as if he thought of them as plots.

Some Remarkable Scatterplots

As Galton's work shows, scatterplots had advantages over earlier graphic forms: the ability to see clusters, patterns, trends, and relations in a cloud of points. Perhaps most importantly, it allowed the addition of visual annotations (point symbols, lines, curves, enclosing contours, etc.) to make those relationships more coherent and tell more nuanced stories. This 2D form of the scatterplot allows these higher-level visual explanations to be placed firmly in the foreground. John Tukey later expressed this as, "The greatest value of a picture is when it forces us to notice what we never expected to see" (1977, p. vi).

In the first half of the twentieth century, data graphics entered the mainstream of science, and the scatterplot soon became an important tool in new discoveries. Two short examples must serve to illustrate applications in physical science and economics.

The Hertzsprung-Russell Diagram

One key feature was the idea that discovery of something interesting could come from the perception—and understanding—of *classifications* of objects based on clusters, groupings, and patterns of similarity, rather than direct relations, linear or nonlinear. Observations shown in a scatterplot could belong to different groups, revealing other laws. The most famous example concerns the Hertzsprung-Russell (HR) diagram, which revolutionized astrophysics.

The original version of the Hertzsprung-Russell diagram, shown here in Figure 6.17, is not a graph of great beauty, but nonetheless it radically changed thinking in astrophysics by showing that scatterplots of measurements of stars could lead to a new understanding of stellar evolution.

Astronomers had long noted that stars varied, not only in brightness (luminosity), but also in color, from blue-white to orange, yellow, and red. But until the early 1900s, they had no general way to classify them or interpret variations in color. In 1905, the Danish astronomer Ejnar Hertzsprung

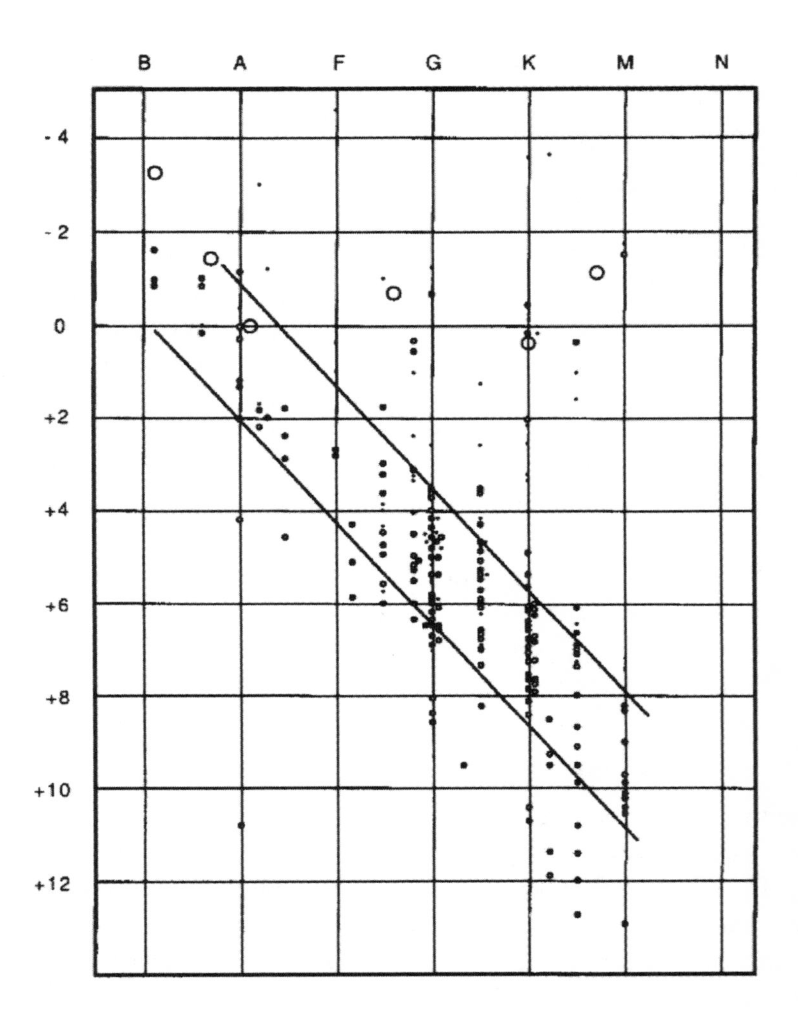

6.17 **Early HR diagram:** Russell's plot of absolute magnitude against spectral class. *Source:* Ian Spence and Robert F. Garrison, "A Remarkable Scatterplot," *The American Statistician*, 47:1 (1993), pp. 12–19, fig. 1. Reprinted by permission of Taylor & Francis Ltd.

presented tables of luminosity and star color. He noted some apparent correlations and trends, but the big picture—an interpretable classification, leading to theory—was lacking, probably because his data were displayed in tables.

This all changed in 1911–1913 when, independently, Hertzsprung and Henry Norris Russell in America prepared scatterplots of luminosity (or

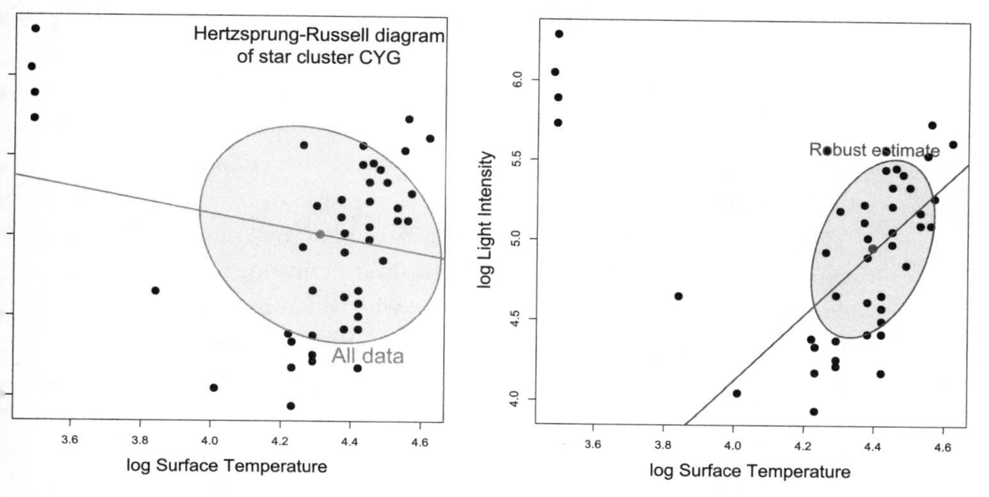

6.18 **Outliers fool least-squares regression:** Hertzsprung–Russell diagrams for a collection of stars in the CYG OB1 star cluster. Left: regression line and data ellipse covering 50% of the points, fit using standard least-squares methods; right: regression line and 50% data ellipse using a robust method which is insensitive to the obvious unusual points in the upper left corner. *Source:* © The Authors.

absolute magnitude) versus the star colors, classified by temperature (or spectral color). They noticed that most of the stars fell along a diagonal band, from the top left (high luminosity, low spectral color) to the low right (low luminosity, high spectral color), which is now called the "main sequence" of stars. They also noticed that other clusters of stars, distinct from the main sequence, were evident. These include what are now called blue and red (super)giants, as well as red and white dwarfs.

The HR diagram showed that stars were concentrated in distinct regions instead of being distributed at random. This regularity was an indication that some definite laws govern stellar structure and stellar evolution. In 1993 Spence and Garrison[27] presented a detailed analysis of the origin, and subsequent development of the HR diagram and its relation to modern statistical graphics. They concluded, "Almost a century after it was first devised, the Hertzsprung-Russell diagram continues to stimulate new directions of inquiry in astronomy" (p. 18).

Figure 6.18 shows modern versions of an HR diagram for stars in the CYG OB1 cluster, which contains forty-seven stars in the direction of Cygnus, recorded by C. Doom.[28] This version plots star brightness against surface temperature (derived from color), both plotted on log scales. Two groups are clearly seen: there is a cluster of four points in the upper left corner (called "giants"); the remaining points (the "main sequence") tend to follow a steep band at the right. The data ellipse is another modern enhancement to the presentation of data in a scatterplot.

This is an aberrant example, where the method of least squares used to fit the line in the left panel is distorted by the four points for the giant stars. It would not have fooled Galton or Herschel, who would have recognized that there was something unusual about the points in the upper left, which perhaps deserved some special explanation. The right panel of Figure 6.18 uses a modern method of *robust estimation* that effectively discounts the deviant points.

The Phillips Curve

In economics, many people had studied changes in inflation, unemployment, import prices, and other variables over time, but it remained most common to plot these as separate series, as Playfair had done (Plate 10). The idea to plot one variable against another did not generally arise in scientific work.

In 1958 New Zealand economist Alban William Phillips published a paper in which he plotted wage inflation *directly* against the rate of unemployment in the United Kingdom from 1861 to 1957. Phillips discovered that, although both variables showed cyclic trends over time, they had a consistent inverse relation. His smoothed curve,[29] shown in Figure 6.19, became one of the most famous curves in economic theory. It became important because economists could understand the *co-variation* in the two variables as representing structural constraints in an economy as a trade-off: to achieve reduced unemployment, the economy must suffer increased inflation (for example, by paying higher wages); to reduce inflation, it must allow more unemployment. With this understanding, policy makers could consider the desired balance between the two.

Figure 6.20 is one of eleven other scatterplots presented in Phillips's paper[30] to illustrate the cyclic nature of inflation and unemployment. This graph also shows why a scatterplot is effective here and time-series plots would not be: the scatterplot shows the inverse relation directly, but the comparison of

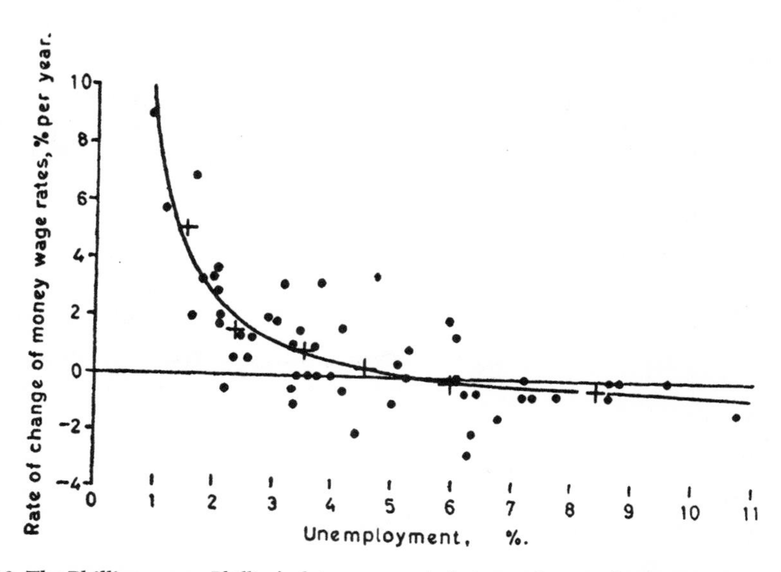

6.19 **The Phillips curve:** Phillips's data on wage inflation and unemployment, 1861–1913, with the fitted Phillips curve. Points used in the fitting process are shown by + signs. *Source:* A. W. Phillips, "The Relation between Unemployment and the Rate of Change of Money Wage Rates in the United Kingdom, 1861–1957," *Economica*, 25:100 (1958), pp. 283–299, fig. 1.

trends over time, as in Playfair's chart of wages and prices (Plate 10), is at best indirect and is subject to the difficulties of using two different vertical scales.

Phillips was not the first economist to use scatterplots, even for time-based data, nor the first to have graphically derived curves named after him.[31] Regardless of priority, Phillips's hand-drawn overall scatterplot (Figure 6.19), combined with his careful parsing of the fitted curve into component cycles (Figure 6.20), provides a final example of Tukey's dictum, another goal scored with a scatterplot.

Spurious Correlations and Causation

As the idea of the scatterplot developed, so too did the mistaken idea that you could plot any variable y against another variable x, and "bingo!" the relationship thus revealed could be interpreted causally. Even though the fallacy, *post hoc ergo propter hoc*, has long been recognized as nonsense, sometimes a causal link seems strengthened with data in a scatterplot.

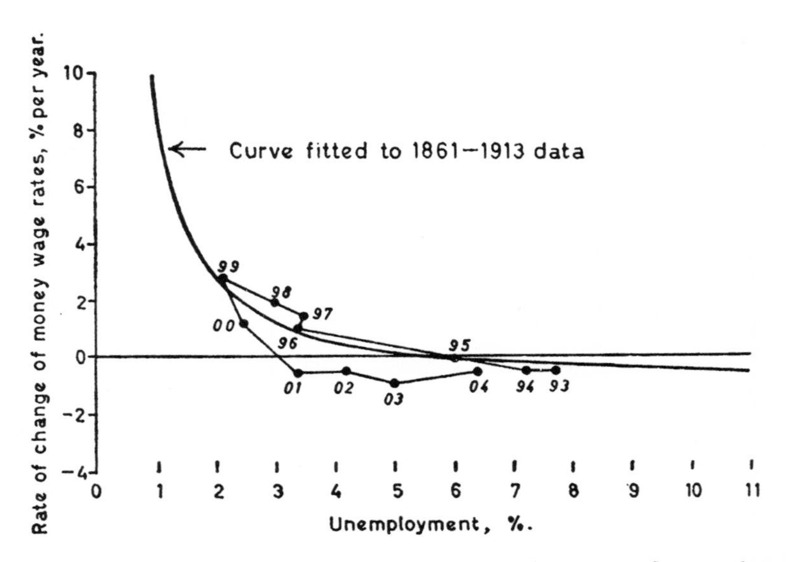

6.20 **Unemployment cycles:** Phillips's data showing one cycle in wage inflation and unemployment, 1893–1904, with the fitted curve from 1861 to 1913. *Source:* A. W. Phillips, "The Relation between Unemployment and the Rate of Change of Money Wage Rates in the United Kingdom, 1861–1957," *Economica*, 25:100 (1958), pp. 283–299, fig. 6.

A 2012 illustration of this was both humorous and subtle.[32] In an article published in the prestigious *New England Journal of Medicine*, Dr. Franz Messerli wondered,[33] "Chocolate consumption could hypothetically improve cognitive function not only in individuals but in whole populations. Could there be a correlation between a country's level of chocolate consumption and its total number of Nobel laureates per capita?" (Messerli, 2012, p. 1562).

The data from twenty-three countries are shown in Figure 6.21. The correlation shown in this plot is $r = 0.79$—not perfect, but suggesting a very strong relationship, which Messerli attributed to the high level of flavanols in chocolate. A popular article in Reuters used the headline, "Eat More Chocolate, Win the Nobel Prize."[34] According to the data, if the average citizen ate only one more kilogram of chocolate per year, their country would gain another 2.5 Nobel prizes.

A rebuttal and an answer of sorts was quickly provided by Pierre Maurage and others in the *Journal of Nutrition* (2013). To test other possible

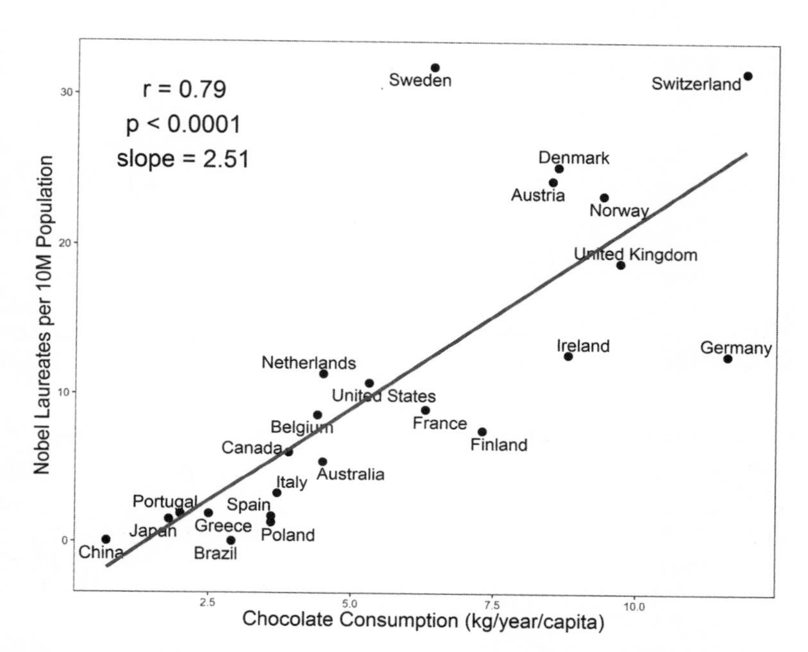

6.21 **A spurious correlation:** Chocolate consumption per capita and the number of Nobel prizes per million population in twenty-three countries. *Source:* Reformatted from Franz H. Messerli, "Chocolate Consumption, Cognitive Function, and Nobel Laureates," *The New England Journal of Medicine*, 367:16 (2012), pp. 1562–1564, fig. 1. © 2012 Massachusetts Medical Society. Reprinted with permission from Massachusetts Medical Society.

explanations for the startling influence of chocolate on Nobel prizes by country, they gathered more data, some of which is shown in Figure 6.22.

A careful scientist tests an explanation by comparing competing hypotheses. If the concentration of flavanols in chocolate is the mechanism, then surely consumption of other flavanol-rich substances—tea and wine—should show a similar strong positive relationship with the number of Nobel prizes. But alas, that did not work out, as can be seen in panel (A) of Figure 6.22.

In a brilliant stroke, they sought an alternative explanation for the distribution of Nobel prizes across countries: the number of IKEA stores per capita. Voilà! As can be plainly seen in panel (B), the number of IKEA stores is an even stronger predictor of Nobel prizes, with $r = 0.82$. Could it be possible that the training required to follow IKEA assembly instructions improves cognitive functioning at the population level even more than chocolate?

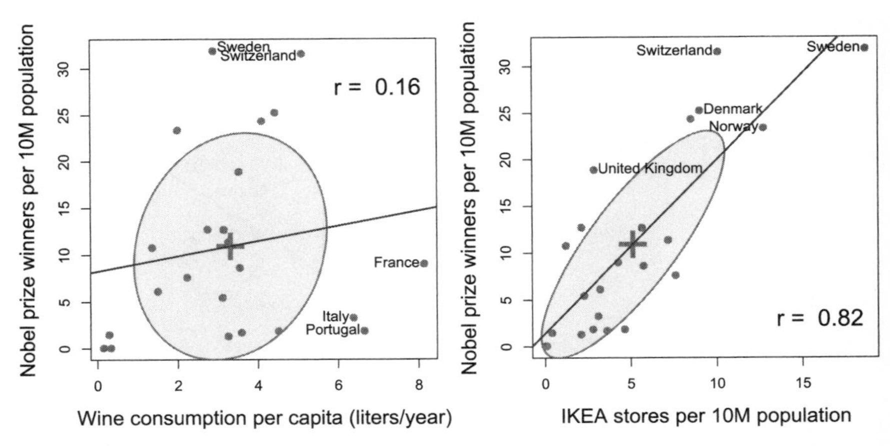

6.22 **More spurious correlations:** Correlations between countries' number of Nobel laureates per 10 million population and other spurious causes: (A) annual per capita wine consumption; (B) number of IKEA stores per 10 million population. *Data source:* Pierre Maurage, Alexandre Heeren, and Mauro Pesenti, "Does Chocolate Consumption Really Boost Nobel Award Chances? The Peril of Over-Interpreting Correlations in Health Studies," *Journal of Nutrition*, 143:6 (2013), pp. 931–933, figs. 1B and 1C.

It took an actual Nobel laureate to unravel the mystery: Eric Cornell, an American physicist who shared the Nobel Prize for physics in 2001, said correctly that "national chocolate consumption is correlated with a country's wealth and high-quality research is correlated with a country's wealth." Correlation does not imply causation: often some other missing third variable is influencing both of the variables you are correlating.

Scatterplot Thinking

Scatterplots took visual displays and analysis beyond one-dimensional problems, passed the idea of plotting functional relations by Halley, and then passed the idea of Playfair's time-series line graphs, where the horizontal axis was bound to time (we call this 1.5 dimensions). The result of the scatterplot was a fully two-dimensional space, where data, depicted by points in a Cartesian framework, were free to roam, constrained only by the relations between the variables, as observed and to be explained.

The need for a scatterplot arose when scientists had to examine bivariate relations between distinct variables directly. As opposed to other graphic forms—pie charts, line graphs, and bar charts—the scatterplot offered a unique advantage: the possibility to discover regularity in empirical data (shown as points) by adding smoothed lines or curves designed to pass "not through, but among them," so as to pass from raw data to a theory-based description, analysis, and understanding (Herschel, 1883b, p. 179).

In the toolbox of modern data graphics, the scatterplot continues to earn its keep, perhaps with a place of pride. Some of the figures in this chapter illustrate the use of modern statistical methods (regression lines, smoothing, data ellipses, and so on) to enhance perception of what should be seen in a cloud of points. It also served as a framework for graphics developers to extend the ideas of Herschel, Galton, and others to higher dimensions and more complex problems.

The advent of computer-generated statistical graphics and software beginning in the 1960s led to other new uses and enhancements. Among these was the perceptually important idea that one could trade off resolution or detail for increased multivariate scope by plotting many smaller scatterplots together in a single, coherent display, in what Tufte (1983) later referred to as "small multiples." One of the first of these new ideas was the idea of a *scatterplot matrix*,[35] a plot of all pairwise relations for p variables in a $p \times p$ grid, where each subplot showed the bivariate relation between the row and column variables.

Other developments would follow this thread: visualizing data in 3D; using motion and dynamic graphics to portray additional features of data varying over time or space; interactive graphics that allow the viewer to query a graph to see more details. These topics are explored in the following chapters.

7

The Golden Age of Statistical Graphics

The period from 1860 to 1890 may be called the golden age of graphics, for it was marked by the unrestrained enthusiasm not only of statisticians but of government and municipal authorities, by the eagerness with which the possibilities and problems of graphic representation were debated and by the graphic displays which became an important adjunct of almost every kind of scientific gathering.

—FUNKHOUSER, 1937, p. 330

With these words, Howard Gray Funkhouser [1898–1984] christened this period in the last half of the nineteenth century as the "golden age of graphics." When he wrote these lines in his PhD thesis at Columbia University in 1937 (which was quickly published in the history journal *Osiris*), he was the first modern writer to attempt a comprehensive history of the graphical representation of statistical data or to see it as a historical topic. As a historian of science, he unearthed the "volumes of forgotten lore" that constituted the early cultivation of this topic, and also established a raison d'être for the study of graphs as scientific objects with an intellectual history.

On many dimensions, this period Funkhouser highlighted as the Golden Age of Graphics was the richest period of innovation and beauty in the entire history of data visualization. During this time there was an incredible development of visual thinking, represented by the work of Charles Joseph Minard, advances in the role of visualization within scientific discovery, as illustrated through Francis Galton, and graphical excellence, embodied in state statistical atlases produced in France and elsewhere.

Ages in the History of Graphics

One convenient way to appreciate the development of ideas and techniques in any field is to record and document the significant events in its history. This is basically what Funkhouser started in his written history of graphical methods.

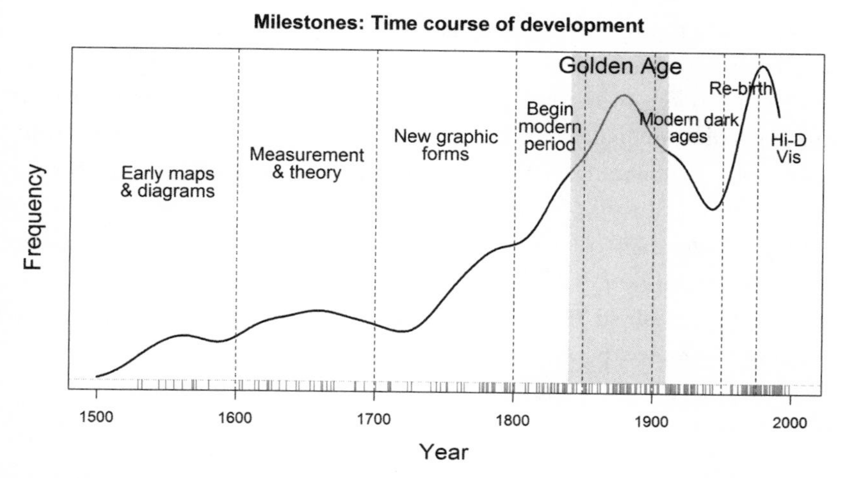

7.1 **Milestones timeline:** The time distribution of events considered milestones in the history of data visualization, shown by a rug plot and a density estimate. The data consist of $n = 260$ significant events from 1500 to the present (Friendly, 2005). The developments in the highlighted period, from roughly 1840 to 1910, comprise the subject of this chapter. *Source:* Reformatted from Michael Friendly, "The Golden Age of Statistical Graphics," *Statistical Science*, 23:4 (2008), pp. 502–535, fig. 1.

The Milestones Project, www.datavis.ca/milestone,[1] does much more. It is a comprehensive online repository for this history, with representative images, references, and text descriptions that can be searched and displayed in various ways and can also analyzed as data on this history.

Figure 7.1 gives a graphic overview, showing the time course of these events from 1500 to the present by a smoothed curve of relative frequency (a kernel density estimate) and fringe marks (a rug plot) at the bottom for the discrete milestone events.

The dashed lines and labels for various periods reflect one convenient parsing of this history.[2] Of interest here is the rapid rise in the early 1800s, which peaked later in this century, followed by a steep decline in the early 1900s, before an even more dramatic rise in the last half of the 1900s.

The first half of the nineteenth century, labeled "Begin modern period" in this graph, is the same historical period described in Chapter 3 as the Age of Data and the time period in which Playfair invented his chart and graphic forms and Dupin, Guerry, and others first used shaded maps to show the geographic distribution of socially important data.

With these innovations in design and technique, the first half of the nineteenth century was also an age of enthusiasm for graphical display.[3] It witnessed explosive growth in statistical graphics and thematic mapping, at a rate that would not be equaled until recent times. This rapid growth continued from about 1840 until about 1900 but did something more than just incremental innovation.

In the latter half of the nineteenth century, this youthful enthusiasm matured, and a variety of developments in statistics, data collection, and technology combined to produce a "perfect storm" for data graphics. As Funkhouser notes, a passion for data graphics became widespread in government agencies and scientific gatherings. The result was a qualitatively distinct period that produced works of unparalleled beauty and scope, the likes of which would be hard to duplicate even today.

A wider perspective can clarify what we think of as an Age and what makes an age Golden. The term *Golden Age* originated from early Greek and Roman poets, who used it to indicate a time when humankind was pure and lived in a utopia; more generally it refers to some recognizable period when great tasks were accomplished: a mountain (or at least a high plateau) of achievement between two valleys. A historical age is a "rise and fall" story; a golden one rises to a spectacular peak.

Some Golden Ages include: (a) the Golden Age of Athens, the forty-four years under Pericles between the end of the Persian War (448 BCE) and the beginning of the Peloponnesian Wars (404 BCE), a relative high point in the development of politics and civil society, architecture, sculpture, and theater; (b) the Golden Age of Islam (750–1258), from the consolidation of the Islamic caliphate to the sack of Baghdad by the Moguls, during which there were great advances in the arts, science, medicine, and mathematics; and (c) the Golden Age of England (1558–1603) under Elizabeth I, a peak in Renaissance literature, poetry, and theater. Such periods often end with one or more turning-point events.

Statisticians might describe a Golden Age as a local maximum in some distribution over history. In Figure 7.1 we can see that the number of milestones events in the history of data visualization grew rapidly throughout the 1800s but then suffered a downturn toward the end of the century. This is just one quantitative indicator of the development of graphical methods in the Golden Age.

Prerequisites for the Golden Age

As we discussed in Chapters 3 and 4, one critical development that launched the invention of the basic forms of statistical graphics in the early part of the nineteenth century was widespread collection of *data* on social problems (crime, suicide, poverty) and disease outbreaks (cholera). In a number of key cases, graphical methods proved their utility, sometimes suggesting explanations or solutions. A second general group of advances that enabled the Golden Age concerned technology, for (a) reproducing and publishing data graphics using color, (b) recording raw data for more than one variable at a time, and (c) tabulating or calculating some summaries that could then be displayed in graphs. A few of these are illustrated in Figure 7.2.

In the period leading up to the Golden Age, thematic maps and diagrams had been printed by copperplate engraving. With this technique, an image is incised on a soft copper sheet, then inked and printed. In the hands of master engravers and printers, copperplate technology could easily accommodate fine lines, small lettering, stippled textures, and so forth. The works of Albrecht Dürer and other engravers attest to how hand-drawn artwork could be transformed into something that captured the artist's intent, with fine lines and texture, and then be printed in many copies. Early data graphic works in

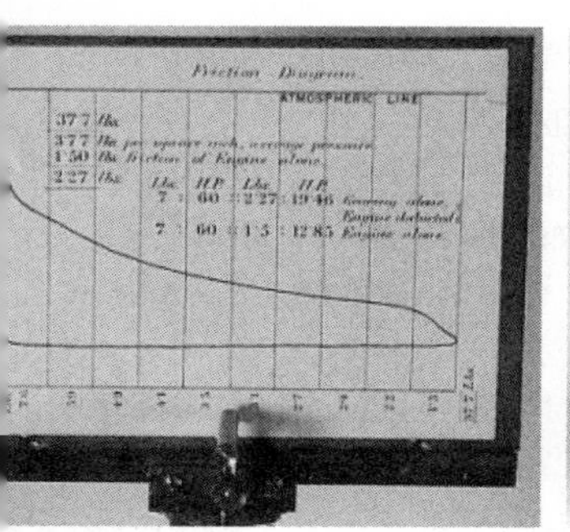

7.2 **Some technological advances leading to the Golden Age:** Left: automatic recording: the Watt Indicator, James Watt (1822); middle: calculating devices, Babbage (1822/1833); right: photography: motion: Muybridge (1879). *Source:* (left) National Museum of American History, The Smithsonian Institution; (middle) Britannica.com; (right) Library of Congress, Prints and Photographs Division, LC-DIG-ppmsca-23778, detail.

this period featured both the author and the engraver in captions or legends, because both had contributed to the final product.

The resulting images were far superior to those produced by previous woodcut methods. But copperplate was slower, more costly, and required different print runs if color was to be used in an overlay inked with a different color. The graphs in Playfair's major works (Playfair, 1786, 1801), for example, were printed via copperplate but hand-colored (often by Playfair himself): hence they were printed in limited numbers.

Lithography, a chemical process for printing invented in 1798 by Aloys Senefelder [1771–1843], allowed much longer print runs of maps and diagrams than engraving, was far less expensive, and also made it easier to achieve fine tonal gradation in filled areas.

By around 1850, lithographic techniques were adapted to color printing, making the use of color less expensive and more frequent. More importantly, color could be more easily used as an important perceptual feature in the design of thematic maps and statistical diagrams; high-resolution color printing is an important characteristic of the Golden Age.[4]

A second major characteristic of the Golden Age was significant advancement in automatic recording. Graphic recording devices—instruments that turn a time-varying phenomenon into a graphic record—date back to antiquity (Hoff and Geddes, 1962). As noted in Chapter 1, Robert Plot constructed a pen device to record the barometric pressure in Oxford every day of 1684, calling the result a "History of the Weather" (Figure 1.4). The basic idea was to find a method, often mechanical, to register some phenomenon and then transfer this to the motion of a pen on a moving roll of paper. New developments in the Golden Age opened a wider range of scientific questions to visual inspection and analysis. Modern seismographs and electroencephalography (EEG) recorders still do much the same, with multiple pens for different channels recorded simultaneously.

By 1822, James Watt (with John Southern) published a description of the "Watt Indicator" (Figure 7.2, left), a device to automatically record the *bivariate* relation between pressure of steam and its volume in a steam engine, with a view to calculating work done and improving efficiency. This remarkable mechanism, which used separate inputs to drive the horizontal and vertical pen positions, let one see directly how these two measures varied together or how one changed as the other was varied.

Such ideas are simple once you see the device. Over the latter half of this century the scope of automatic recording expanded enormously, going from weather and physical measurement to questions of the flight of birds and the physiology of the human body. A key player in this technology was Étienne-Jules Marey, whose contributions are detailed in Chapter 9.

A third significant advancement in this period was in calculation. The wealth of data collected in the early nineteenth century created a need for serious number crunching, to summarize and make sense of them. A large number of mechanical calculating devices were developed earlier in the seventeenth century to meet this need, providing the rudiments of four-function calculators (addition, subtraction, multiplication, and division).[5] This changed, at least in theory, in 1822 when Charles Babbage [1791–1871] conceived the "Difference Engine," a mechanical device for calculating mathematical tables of logarithms and trigonometric functions and automatically printing the results; somewhat later (1837), he designed the "Analytical Engine," a mechanical general-purpose and programmable computing device that received program instructions and data via punched cards (such as had been used on Joseph Jacquard's mechanical looms to program the sequence of colors in weaving). Ada Lovelace recognized the immense potential of Babbage's device to be programmed to "weave algebraic patterns" and in 1833 invented what many consider to be the first computer program (to calculate Bernoulli numbers). Neither of these was actually constructed in Babbage's lifetime, but the idea of tabulating large volumes of data was in the air throughout the Golden Age.

The first-known actual device for tabulating large-scale data is attributed to Andre-Michel Guerry in conjunction with his 1864 work on crime and suicide in England and France over twenty-five years. His data included 226,000 cases of personal crime classified by age, sex, and other factors related to the accused and the crime, and 85,000 suicides, classified by motive. He invented an apparatus, the *ordonnateur statistique*, to aid in the analysis and tabulation of these numbers.[6]

By 1890, in time for the decennial US Census, Herman Hollerith [1860–1929] introduced a modern form of punched card to store numerical information, a keypunch device for entering data, and mechanical devices for counting and sorting the cards by columns of data. After an operator placed punched cards in the hopper, selected a column, and pressed the "Start" button, the cards were then sorted and counted by age group, occupation, religion, or anything else that had been recorded. Answers to questions came out in the number of cards appearing in each bin and recorded by numbers on dials.

Lastly, a final aspect of the graphic language that contributed to the Golden Age, albeit indirectly, arose from the practical needs of civil and military engineers for easy means to perform complex calculations without anything more than a calculating diagram (or "nomogram"), a straightedge, and a pencil. For example, artillery and naval engineers created diagrams and graphical tables for calibrating the range of their guns. Léon Lalanne, an engineer at the Ponts et Chausées, created diagrams for calculating the smallest amount of earth that had to be moved when building railway lines in order to make the work time- and cost-efficient (Hankins, 1999).

Perhaps the most remarkable of these nomograms was Lalanne's (1844) "Universal calculator," which allowed graphic calculation of over 60 functions of arithmetic (log, square root), trigonometry (sine, cosine), geometry (area, circumference and surface of geometrical forms), and conversion factors among units of measure and practical mechanics.[7] In effect, Lalanne had combined the use of parallel, nonlinear scales such as those found on a slide rule (angles to sine and cosine) with a log-log grid on which any three-variable multiplicative relation could be represented by straight lines. For the engineer, it replaced books containing many tables of numerical values. For statistical graphics, it anticipated ideas of scales and linearization used today to simplify otherwise complex graphical displays.

We illustrate this slice of the Golden Age with Figure 7.3, a tour-de-force graphic by Charles Lallemand (1885) for precise determination of magnetic deviation of the compass at sea in relation to latitude and longitude. This multifunction nomogram combines many variables into a device for graphic calculation through complex trigonometric formulas represented visually. It starts at the left with a map of the navigable world from Europe to the Americas, but deformed into what is called an "anamorphic" map, so that the dark lines shown for magnetic declination have a more consistent pattern than on a standard map. It also incorporates 3D figures, parallel coordinates, and hexagonal grids. In using this device, the mariner plots his position at sea on the anamorphic map at the left, projecting that point through the upper central cone, then onto the grids and anamorphic maps at the right, and finally

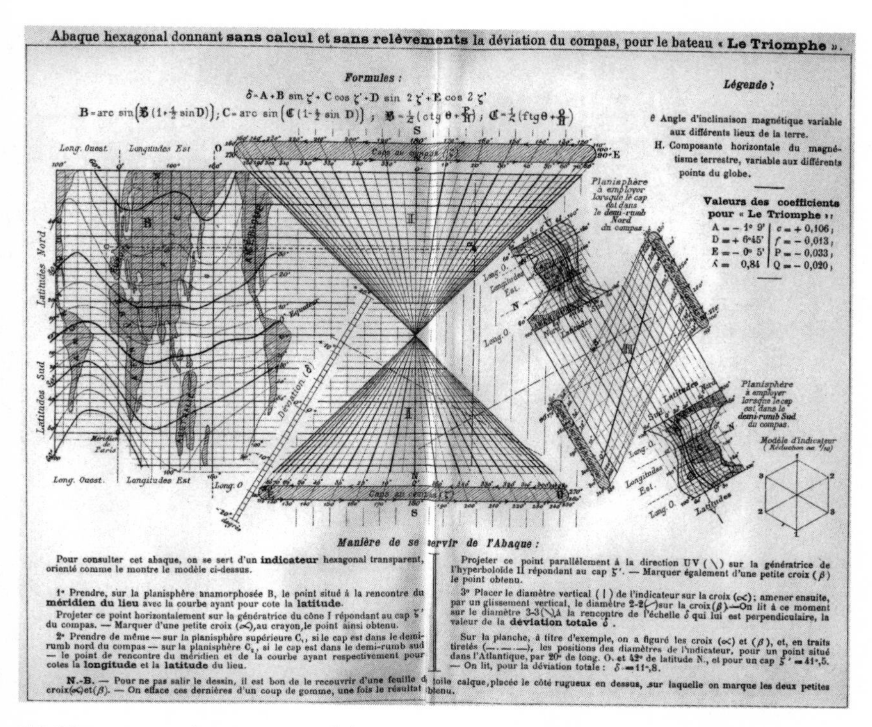

7.3 **Nomograms:** A computational diagram combining diverse graphic forms. This tour-de force nomogram by Charles Lallemand (1885) uses anamorphic maps, parallel coordinates, and 3D surfaces to calculate magnetic deviation at sea. *Source:* Reproduction courtesy of École des Mines.

through the bottom central cone onto the scale of magnetic deviation. Voila! The mariner can assure the crew they'll be home in time for Sunday dinner.

The Graphic Vision of Charles Joseph Minard

The dominant principle which characterizes my graphic tables and my figurative maps is to make immediately appreciable to the eye, as much as possible, the proportions of numeric results. . . . Not only do my maps speak, but even more, they count, they calculate by the eye.

—MINARD (1862b)

Charles Joseph Minard [1781–1870] is most widely known for his compelling portrayal of the terrible losses suffered by Napoleon's Grand Army in the

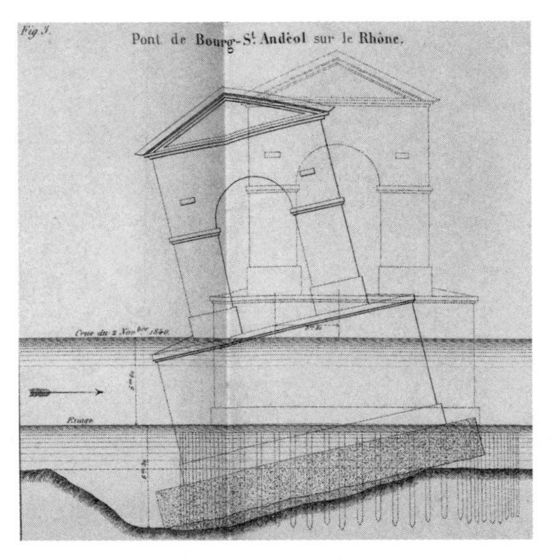

7.4 **Engineering diagram:** Why did the bridge collapse? A cross-sectional diagram showing one of the bridge supports, providing a before-after comparison. *Source:* Detail from Charles Joseph Minard, *De la chute des ponts dans les grandes crues.* Paris: E Thunot et Cie, 1856. Reproduction © École nationale des ponts et chausées, 4_4921_C282.

1812 campaign on Moscow. Edward Tufte (1983) called it "the best statistical graphic ever produced" (p. 40). However, Minard's wider work serves to illustrate the rise of visual thinking and visual explanation that began in the early nineteenth century and came to fruition in the Golden Age.

Minard was trained as an engineer at the École Nationale des Ponts et Chaussées (ENPC), the prestigious French National School of Bridges and Roads. He had two distinct careers there: he first served (1810–1842) as a civil engineer, designing plans for construction of canals and railways, and afterward (1843–1869) he worked as what can be called a visual engineer for the modern French state.

Figure 7.4 shows an example of visual thinking and visual explanation from his early career. In 1840, Minard was sent to Bourg-Saint-Andèol to report on the collapse of a suspension bridge across the Rhône, constructed only ten years before and therefore a major embarrassment for the ENPC. Minard's findings consisted essentially of this self-explaining before-and-after diagram. The visual message was immediate and transparent: apparently, the riverbed beneath the supports on the upstream side had eroded, leaving the bridge unsupported over a good portion of its width. His 1856 pamphlet contains other similar engineering visual explanations.

Minard produced sixty-three known graphic works in the 1843–1869 period.[8] These included *tableaux graphiques* (charts and statistical diagrams) and *cartes figuratives* (thematic maps). Before his retirement in 1851 his "bread-and-butter" topics concerned matters of trade, commerce, and transportation: Where to build railroads and canals? How to charge for transport of goods and passengers? How to visualize changes over time and differences over space? Most of his thematic maps were flow maps, which he developed to a near art form. His choice of the term *carte figurative* signals that the primary goal was to represent the data; the map was often secondary.

Minard's graphic vision, with the primary goal of representing the data, is vividly seen in the pair of before-and-after flow maps in Figure 7.5. His goal was to explain the effect that the US Civil War had on trade in cotton between Europe and elsewhere. Again, the visual explanation is immediate and interocular—it hits you between the eyes. In 1858, most of the cotton imported to Europe came from the US southern states—the wide dark band (blue in the original—Plate 17) that dominates the left figure. By 1862, the blockade of shipping to and from the South reduced this supply to a trickle, which came entirely through the port at New Orleans; some of the demand was met by Egyptian and Brazilian cotton, but the bulk of the replacement came from India. In order to accommodate the flow lines, he widened the English Channel and the Strait of Gibraltar; to make the data stand out; he reduced the coastline of North America to a mere cartoon form.

As mentioned previously, Minard's greatest work, lionized as "perhaps the best statistical graphic ever produced"[9] (Figure 7.6), that portrays the catastrophic loss of life by Napoleon's Grand Army in his ill-fated 1812 invasion of Russia. You could just look at it and think "Oh, that's sort of nice." But, *no*—this is an epic story, told in a single graph.

The lighter flow line begins at the Niémen River at the Polish border on the left where Napoleon began his invasion on June 24, 1812. The width of the flow line reflects the size of the army, initially 422,000 strong (including conscripts from his empire), and its path shows the route taken. Key battles and events along the way are shown on a schematic map of Russia, and we see how his army diminished in strength as it approached Moscow at the right.

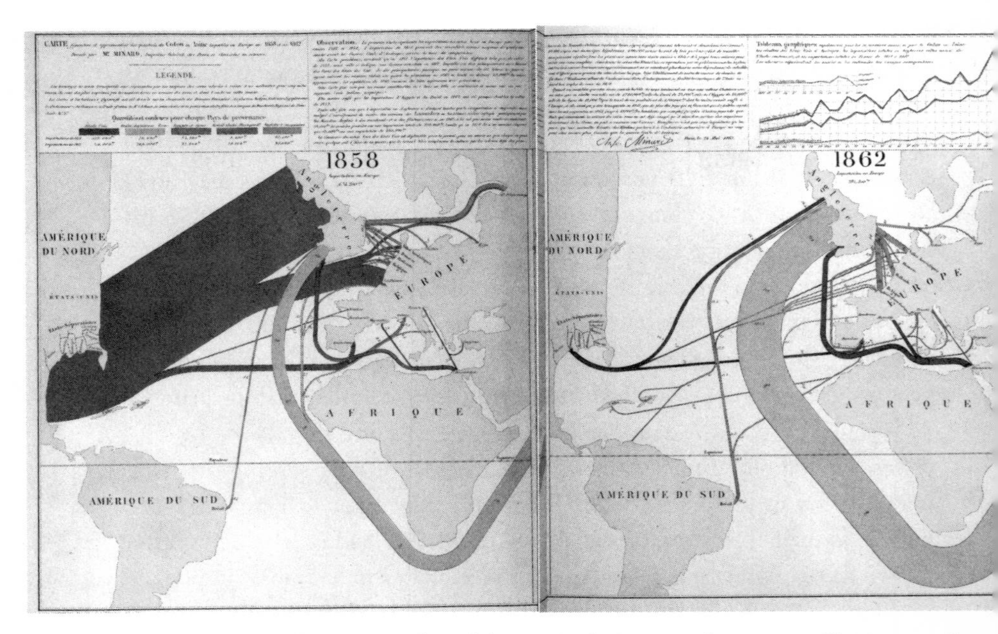

7.5 **Comparative flow maps:** Effect of the US Civil War on trade in cotton. The import of raw cotton to Europe from various sources to destination is shown by colored flow bands of width proportional to the amount of cotton before (left: 1858) and after (right: 1862) the US Civil War. *Source:* Charles Joseph Minard, "Carte figurative et approximative des quantités de coton en Europe en 1858 et 1862," Paris, 1863. Reproduction © École nationale des ponts et chausées, 4Fol 10975.

Tchaikovsky's *1812 Overture* celebrates of the defense of Moscow by a much smaller, underequipped army against this previously undefeated foe.[10] Tolstoy's *War and Peace*, notorious as among the longest novels ever written, also conveys a view of this history from the Russian side.

But Minard, in but a single graph, told his French story of Napoleon's defeat. This is something that, as far as we know, had never been done before (or maybe since) in a graphic portrayal of the history of one's own country. It is a patriot's graphic story, a sad reflection on the folly of war for military conquest.

The path of Napoleon's retreat is shown by the black flow line, vastly diminished, but still 100,000 strong leaving Moscow. By the time they returned to the Niémen river, only 10,000 (about 2 percent) were left. The subscripted

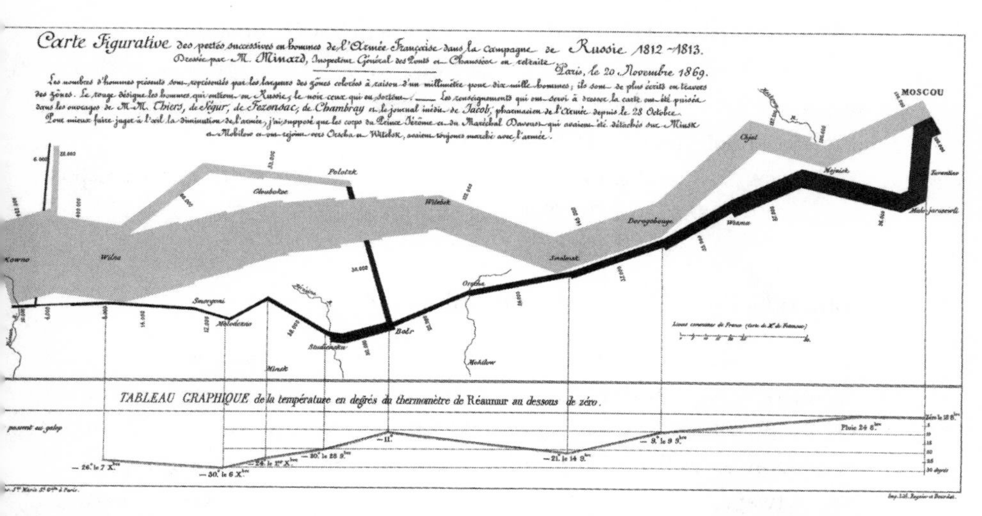

7.6 **Minard's greatest work:** Minard's 1869 *Carte figurative* depicting the fate of Napoleon's Grand Army in the disastrous 1812 campaign to capture Moscow. *Source:* Wikimedia Commons.

graph at the bottom of the chart tries to tell why: Napoleon began his retreat on October 19, 1812, and his supplies were largely exhausted. As the army struggled back, the graph of the declining temperature over the Russian winter symbolizes the brutal conditions that accompanied the soldiers on their terrible retreat.

Minard's works were printed in limited numbers, and he was not well known outside the small circle of French engineers and those interested in the graphic method. Étienne-Jules Marey (1878) first called attention to this powerful graphic, which might otherwise have been lost to history. In the first general book on the graphical method, he said Minard's work "defies the pen of the historian in its brutal eloquence." Later, Funkhouser (1937), in the first modern overview of graphical methods, devoted several pages to Minard's work and called him "the Playfair of France," to suggest the scope of his contributions. Tufte (1983) also brought this image to wide popular attention, describing it as showing "multivariate complexity integrated so gently that viewers are hardly aware that they are looking into a world of six dimensions. . . . It may well be the best statistical graphic ever produced" (p. 40).

Minard died in 1870, and the well-known March on Moscow graphic (published November 20, 1869) was, along with a similar graphic of Hannibal's army in Italy, among his last published works. We return to a wider appreciation of Minard in Chapter 10.

We recently discovered Minard's burial site in Montparnasse Cemetery, Section 7, 48.8388° N, 2.3252° E (Friendly et al. 2020).

Francis Galton's Greatest Graphical Discovery

In Chapter 6, we learned of Galton's justly famous 1886 discovery of the idea of regression and the concentric ellipses that characterize the bivariate normal distribution, which, a decade later, led to Karl Pearson's theory of correlation. However, Galton had achieved an even more notable graphic discovery twenty-five years earlier, in 1863—uncovering the relation between barometric pressure and wind direction that now forms the basis of modern understanding of weather. Most notably, this is a shining example of a scientific discovery achieved almost entirely through graphical means, "something that was totally unexpected, and purely the product of his high-dimensional graphs."[11]

Galton, a true polymath, developed an interest in meteorology around 1858, after he was appointed a director of the observatory at Kew. This work suggested many scientific questions related to geodesy, astronomy, and meteorology; but in his mind, any answers depended first on systematic and reliable data, and second on the ability to find coherent patterns in the data that could contribute to a general understanding of the forces at play.

In 1861 Galton began a crowd-sourced campaign to gather meteorological data from weather stations, lighthouses, and observatories across Europe, enlisting the aid of over 300 observers. His instructions included a data collection form (Figure 7.7) to be filled out at 9 AM, 3 PM, and 9 PM, for the entire month of December 1861, recording barometric pressure, temperature, wind direction and speed, and so forth. From the returns, he began a process of graphical abstraction, which was eventually published in 1863 as *Meterographica*. Altogether, he made over 600 maps and diagrams, using lithography and photography in the process. In his program Galton had a collection of standardized data across all of Europe to make the recordings of these seven variables comparable; a keen appreciation of the power of graphical methods

Contributors, according to the Conditions of my Circular Letter, are requested to enter their Observations in one of the blank forms, to enclose it in a stamped envelope, and to post it to my address on January 1st, 1862.

FRANCIS GALTON,
42, *Rutland Gate, London.*

Name of Station :
Its Latitude :
Its Longitude from Greenwich :
Its Height above Sea Level, in English Feet :

Name of Contributor :
Full Address to which the Charts are to be forwarded when ready :

Date. Either Local or Railway Time; state which. December 1861.		Barometer corrected to Freezing Point at Mean Sea Level, and reduced to English Inches, Tenths, and Hundredths.	Exposed Thermometer in Shade, to nearest Degree, Fahrenheit.	Moistened Bulb to nearest Degree, Fahrenheit, for Evaporation and Dew Point.	Direction of Wind, *true* not magnetic. Only 16 points of the Compass are used; as, N., N.N.E., N.E., E.N.E., E., &c.	Force of Wind: Calm, Gentle, Moderate, Strong, Gale.	Amount of Cloud: Clear blue sky, A few clouds, Half clouded, Mostly clouded, Entirely clouded, Entirely and heavily clouded.	Rain, Snow, or neither.	REMARKS.
1	*9* A.M.								
	3 P.M.								
	9 P.M.								
2	*9* A.M.								
	3 P.M.								
	9 P.M.								
3	*9* A.M.								
	3 P.M.								
	9 P.M.								

7.7 **Galton's data collection form:** Top portion of the form Galton sent to observers to record weather variables throughout the month of December 1861. Noteworthy is Galton's attempt to define the conditions of the observations and to standardize the scales on which each of the seven weather variables were to be recorded. *Source:* Francis Galton, *Meteorographica, or Methods of Mapping the Weather*. London: Macmillan, 1863.

to reveal systematic patterns; and an ability to invent or adapt visual symbols for his purpose.

In the first stage, Galton constructed ninety-three maps (three per day, for each of thirty-one days) on which he recorded multivariate glyphs using stamps or templates he had devised to show rain, cloud cover, and the direction and force of the wind, as shown in Plate 11.[12] He explained that these visual symbols were just as precise as the letters N, NNW, NW, and so on to express wind direction, but the icons "have the advantage of telling their tale directly to the eye" (p. 4).

Although these maps showed all of the data visually, they gave far too much information to see general patterns, particularly when spread across ninety-three pages. He needed a way to compress and summarize the data to capture systematic variation over both time and space. He hit upon the idea of making iconic maps on a geographical grid, to show barometric pressure (Figure 7.8).

Then Galton saw something striking. At this time, a theory of cyclones suggested that in an area of low barometric pressure, winds spiraled inward, rotating counterclockwise. Galton was able to confirm this from his charts, but

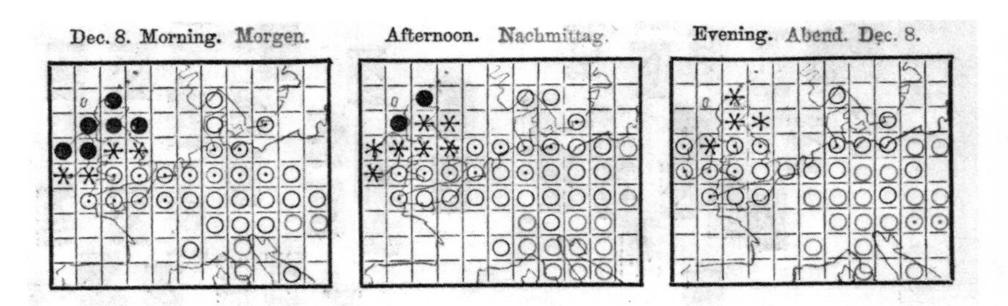

7.8 **Iconic 3D barometric maps, bipolar scale:** Galton's barometric maps for December 8, 1861. Gray and black symbols represent, respectively, lower and higher barometric pressure than average, with degrees of divergence ranging from ○ through ⊙, and ✻ to ●. *Source:* Francis Galton, *Meteorographica, or Methods of Mapping the Weather*. London: Macmillan, 1863.

he noticed something else. Across geographic space, areas of high barometric pressure also corresponded to an *outward* spiral of wind in the clockwise direction, a relation he called an "anticyclone" (Galton, 1863a). This observation formed the basis for a more general theory of weather patterns, linking barometric pressure to wind and other weather variables.

What would prove key to confirming this idea was an ability to see *relationships* of wind direction and pressure over space and *changes* in them over time at a more global level. Therefore, in a second stage of abstraction, he reduced the data for each day to a 3 × 3 grid of miniature abstract contour maps. In the rows, these showed mini-maps of barometric pressure, wind direction and rain, and temperature; the columns represented morning, noon, and afternoon. In these mini-maps Galton used color, shading, and contours to show approximate iso-levels and boundaries, and arrows to show wind direction.

He assembled these mini-maps for all thirty-one days into a single two-page chart of multivariate schematic mini-maps, of which the right-hand page is shown in Plate 12. The legend for the symbols appears in the left panel of Plate 13. The portion representing the data for December 5 is shown in the right panel.

Conveniently, it turned out that barometric pressure (in the top row for each day) was generally low in the first half of December and high in the second half. The correlated directions of the arrows for wind direction confirmed the theory. He explained these results with reference to Dove's Law of Gyration (Galton, 1863a). A prediction from this and Galton's cyclone-anticyclone theory was that a reversed pattern of flow should occur in the southern hemisphere; this was later confirmed.

Galton's discovery of weather patterns illustrates the combination of complex data with visual thinking. It also illustrates the considerable labor to simplify the data to highlight patterns and finally to produce a theoretical description. His further work in meteorology also illustrates the translation of theory into practical application, another feature we find prominent in the Golden Age.

From 1861 to 1877, he published seventeen articles dealing with meteorological topics, such as how charts of wind direction and intensity could be translated into charts of travel time for mariners.[13] On April 1, 1875, the London *Times* published a weather chart prepared by Galton; this was the first instance of the modern weather maps we see today in newspapers worldwide.[14]

Statistical Albums

A final exemplar of the Golden Age was a collection of government projects of unparalleled graphical excellence, scope, and beauty. The collection, organization, and dissemination of official government statistics on population, trade and commerce, and social and political issues became widespread in most European countries from about 1820 to 1870. After about 1870, the enthusiasm for graphic representation took hold in many of the state statistical bureaus in Europe and the United States, resulting in the preparation of a large number of statistical atlases and albums. As befits state agencies, the statistical content and presentation goals of these albums varied widely, and the subject matter was often mundane, but the results were spectacular, even today.

In the United States, the US Census Bureau, under the direction of Francis Walker, produced statistical atlases to depict the demographic characteristics of the population by age, gender, religion, and national origin, but occasionally some wider topics: manufacturing and resources, taxation, poverty, and crime. In France, the Ministry of Public Works focused largely on aspects of trade, commerce, and transportation.[15]

Regardless of their content, the resulting publications are impressive, for their wide range of graphic methods and often for the great skill of visual design they reflect. As we shall see, they often anticipated graphical forms and ideas that were only reinvented after 1970.

The Album de Statistique Graphique

Minard's graphic works at the ENPC were very influential in government bureaus in France, so much so that nearly every minister in the Ministry of Public Works from 1850 to 1860 had his official portrait painted with one of Minard's works in the background.[16] In March 1878, the ministry established a bureau of statistical graphics under the direction of Émile Cheysson [1836–1910]. Like Minard, Cheysson had been an engineer at the ENPC until his appointment to the Ministry of Public Works. He was the major representative of France in committees on the standardization of graphical methods at the International Statistical Congresses from 1872 on.

By July 1878, the new bureau was given its marching orders and charged to "prepare *(figurative)* maps and diagrams expressing in graphic form statistical documents relating to the flow of passenger travel and freight on lines of communication of any kind and at the seaports, and to the construction and exploitation of these lines and ports; in sum, all the economic facts, technical or financial, which relate to statistics and may be of interest to the administration of public works."[17]

From 1879 to 1897 the statistical bureau published the *Album de Statistique Graphique*. These volumes were large-format quarto books (about 11×15 in.), and many of the plates folded out to four or six times that size; all plates were printed in color and with great attention to layout and composition. Funkhouser noted (1937, p. 336) that "the *Albums* present the finest specimens of French graphic work in the century and considerable pride was taken in them by the French people, statisticians and laymen alike." It is no stretch to claim these volumes as the pinnacle of the Golden Age, an exquisite sampler of nearly all known graphical forms, and a few that made their first appearance in these volumes.[18]

These albums had two general themes: the main topics concerned economic and financial data related to the planning, development and administration of public works—transport of passengers and freight, by rail, on inland waterways and through seaports; imports, exports, and expenditures on infrastructure. In addition, occasional topics, which varied from year to

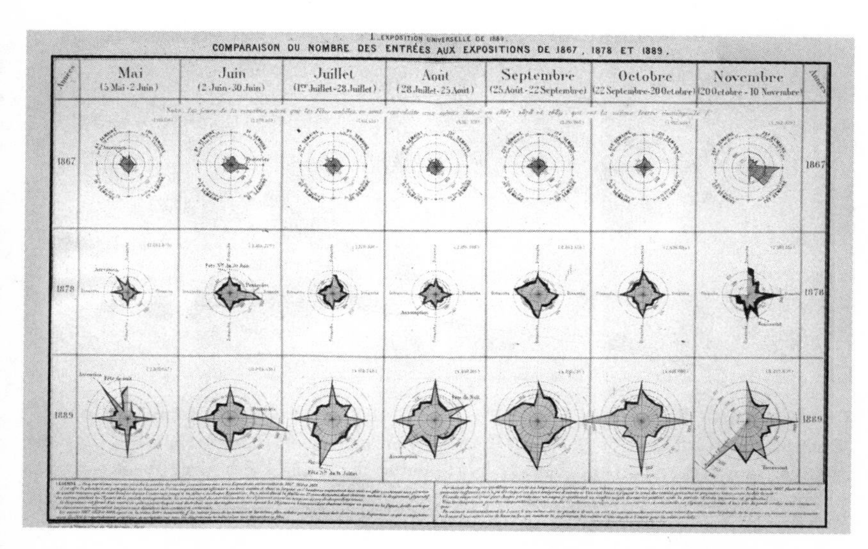

7.9 **Two-way star / radar diagrams:** "Comparison of the numbers attending the Expositions of 1867, 1878 and 1889" (*Exposition Universelle de 1889: Comparaison du Nombre des Entrées aux Expositions de 1867, 1878 et 1889*). In each star-shaped figure the length of the radial dimension shows the number of paid entrants on each day of the month. *Source:* Caisse nationale des retraites pour la vieillesse. *Album de Statistique Graphique*. Paris: Imprimerie Nationale, 1889, Plate 21.

year, included such subjects as agriculture, population growth, transport, international expositions in Paris, and so forth. The first theme was the raison d'être of the bureau; the second allowed Cheysson and his team to delight their readers with something new, relevant to a topic of interest, often using some novel graphic design. The menu for ministers and officials who received these reports was clear: bread and butter were served with a hardy main course of visualized statistics, followed by eye-candy for dessert.

The 1889 volume followed that year's universal exposition in Paris, and it used several novel graphic designs to provide an analysis of data related to this topic. Figure 7.9 uses what are now called *star* or *radar diagrams* to show attendance at each of the universal expositions held in Paris: 1867, 1878, and 1889. These are laid out as a two-way array of plots, in a form we now call a "trellis display,"[19] to allow comparisons of the rows (years) and columns (months). Each star diagram shows daily attendance by the length of the ray, in yellow for paid entrance and black for free admissions, with Sundays oriented at the

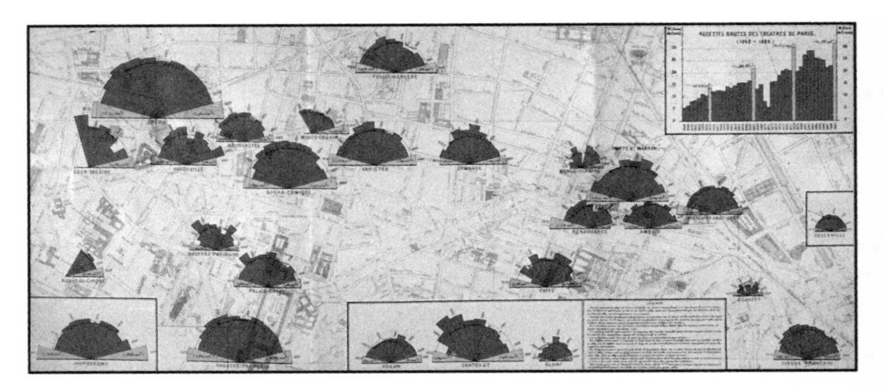

7.10 **Polar / area diagrams on a map:** "Gross receipts of theaters in Paris from 1878 to 1889" (*Exposition Universelle de 1889: Recettes brutes des théatres et spectacles de Paris 1878 à 1889*). Each diagram uses sectors of length proportional to the receipts at a given theater in each year from 1878 to 1889, highlighting the values for the years of the Universal Expositions in a lighter shade. *Source:* Caisse nationale des retraites pour la vieillesse. *Album de Statistique Graphique*. Paris: Imprimerie Nationale, 1889, Plate 26.

compass points. In this display, we can see: (a) attendance increased greatly from 1867 to 1889; (b) Sundays were usually most well-attended; and (c) in 1889, there were a number of additional spikes, mostly holidays and festivals, which are noted on the graphs with textual descriptions.

Another graphic in the same volume tried to highlight the impact of these expositions on attendance in theaters in Paris. Figure 7.10 shows polar diagrams for the major theaters, with the area of each sector proportional to gross receipts in the years from 1878 to 1889. The expo years of 1878 and 1889 are shaded yellow, and others are shaded red. These figures are placed on a map of the right bank of Paris, with theaters elsewhere shown in boxes. This image inventively combines polar area charts with a faint background map of Paris to provide geographical context. The histogram in the upper right corner shows total receipts for all years from 1848 to 1889.

Figure 7.11 is yet another singular plate, from the 1888 Album. It uses what is called an *anamorphic map* to show how travel time in France (from Paris) had decreased over two hundred years. Cheysson's graphic idea, which was far ahead of its time, is simple: shrink the map to make travel time in different years proportional to distance in the map. Here the outer boundary of the map represents, along each radial line, the travel time to various cities in 1650.

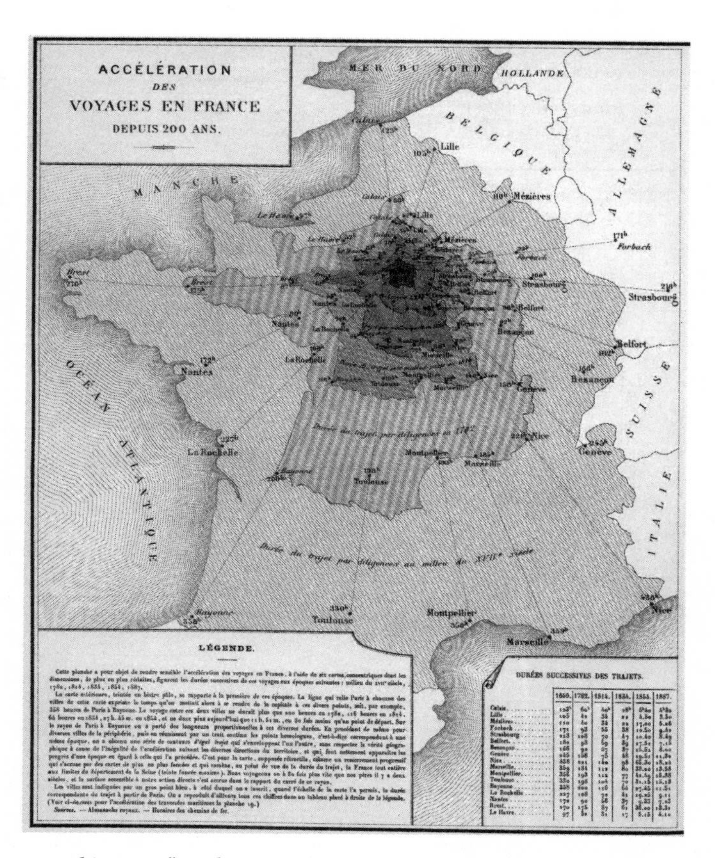

7.11 **Anamorphic map:** "Acceleration of travel in France over 200 years" (*Accélération des voyages en France depuis 200 Ans*). A set of five Paris-centric maps scaled along radial directions to major cities to show the relative decrease in travel time from 1789 to 1887. *Source:* Caisse nationale des retraites pour la vieillesse. *Album de Statistique Graphique.* Paris: Imprimerie Nationale, 1888, Plate 8a.

These lines are then scaled in proportion to the reduced travel time in the years 1789, 1814, and up to 1887, with the numerical values shown in the table at the bottom right and along each radial line.[20] The outline of the map of France was then scaled proportionally along those radial lines. What becomes immediately obvious is that the shrinking of travel times was not uniform. For example, travel time to the north of France (Calais, Lille) decreased relatively quickly; in the south, Montpelier and Marseilles "moved" relatively closer to Paris than did Nice or Bayonne in this period.

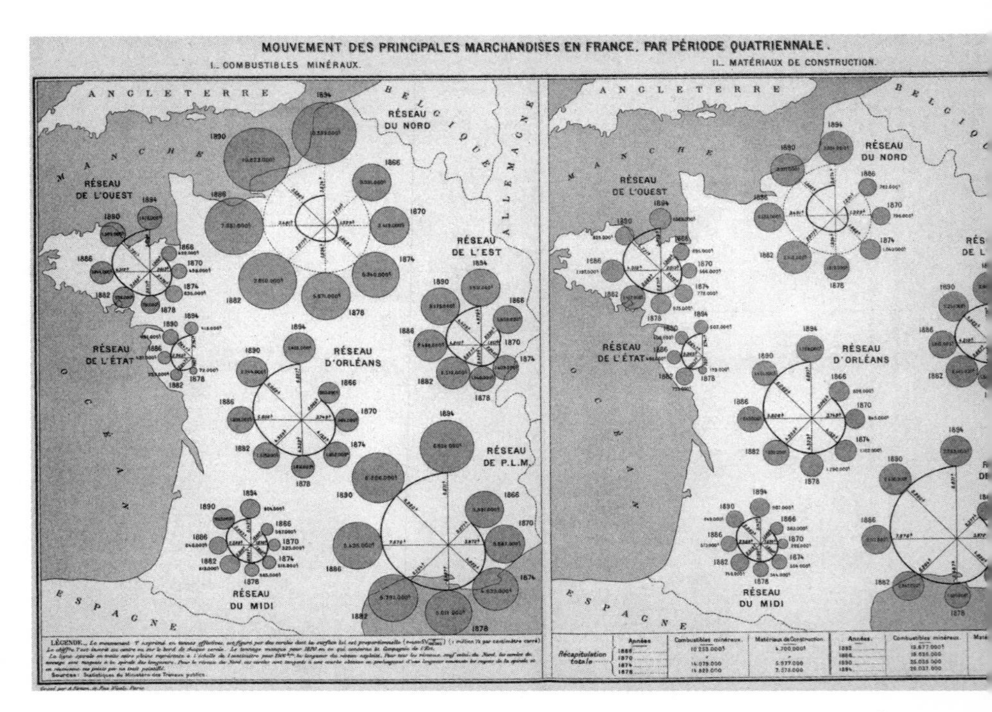

7.12 **Planetary diagram:** "Transportation of principal merchandise in France in four-year periods" (*Mouvement des principales marchandises en France, par période quatriennale*). Left: combustible minerals, for example, coal, coke; right: construction materials. The length of rays indicate average distance; circle diameters represent tonnage moved. *Source:* Caisse nationale des retraites pour la vieillesse. *Album de Statistique Graphique*. Paris: Imprimerie Nationale, 1897, Plate 9.

This graphic form is now more generally called a *cartogram*: some thematic mapping variable, such as travel time, population, rates of HIV infection, or votes for a political party, is substituted for land area or distance and the geometry of the map is distorted to convey that information directly. Cartograms of various forms now provide a powerful way to blend data into a map, giving the data prominence.[21]

One challenge Cheysson faced was how to show changes over time for two or more related variables simultaneously in relation to the geography of France. The Albums used a variety of novel graphic forms for this purpose. For example, Figure 7.12 uses "planetary diagrams" to show two time series of the transportation of principal merchandise by region over the years

1866–1894 in four-year intervals. The rays of the spiral are proportional to the average distance traveled; the diameters of the circles are proportional to tonnage moved.

US Census Atlases

Other striking examples representing high points of the Golden Age appear in the series of statistical atlases published by the US Census Bureau in three volumes for the decennial census years 1870 to 1890. The *Statistical Atlas of the Ninth Census*, published in 1874 under the direction of Francis A. Walker [1840–1897] was the first true US national statistical atlas, composed as a graphic portrait of the nation. This was followed by larger volumes from each of the 1880 and 1890 censuses, prepared under the direction of Henry Gannett [1846–1914], sometimes described as the father of American government map-making.[22]

The impetus for this development stemmed largely from the expanded role given to the census office following the US Civil War. The decennial census, which was begun in 1790 by Thomas Jefferson, was initially designed to serve the constitutional need to apportion congressional representation among the states. However, by June 1872, the Congress recognized "the importance of graphically illustrating the three quarto volumes of the ninth census of the United States, by a series of maps exhibiting to the eye the varying intensity of settlement over the area of the country, the distribution among the several States . . . , the location of the great manufacturing and mining industries, the range of cultivation of each of the staple productions of agriculture, the prevalence of particular forms of disease and other facts of material and social importance which have been obtained through such census."[23]

Accordingly, the atlas for the ninth census was composed of fifty-four numbered plates divided into three parts: (a) physical features of the United States: river systems, woodland distribution, weather, minerals; (b) population, social, and industrial statistics: population density, ethnic and racial distribution, illiteracy, wealth, church affiliation, taxation, crop production, and so on; (c) vital statistics: age, sex, and ethnicity distributions, death rates by age, sex, causes, distributions of the "afflicted classes" (blind, deaf, insane), and so on. The plates were accompanied by eleven brief discussions of these topics, containing tables and other illustrations.

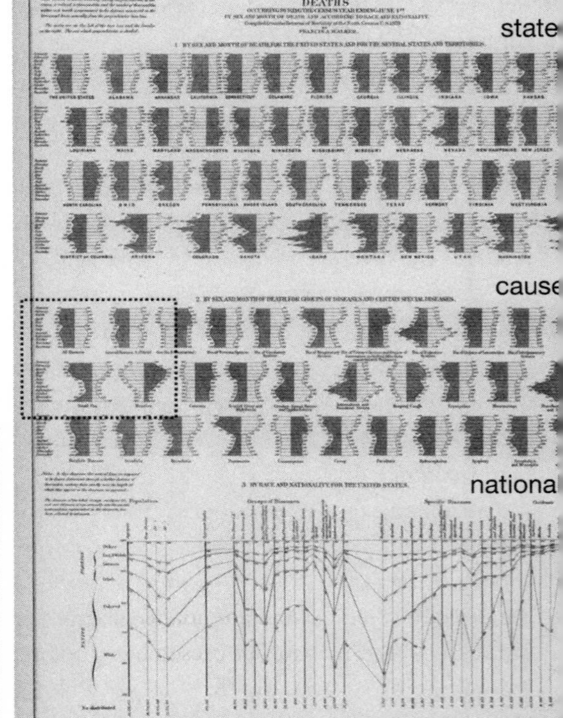

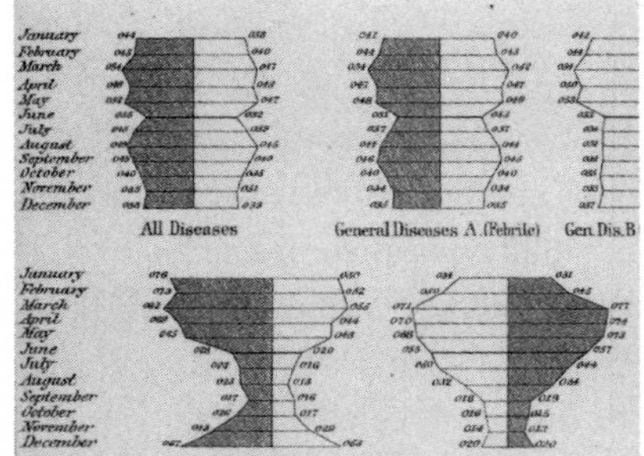

7.13 **Bilateral histograms:** "Chart Showing the Distribution of Deaths . . . by Sex and Month of Death and according to Race and Nationality." Left: detail from causes of death; right: full plate, with labels for the three sections added. *Source:* United States Census Office, *Statistical Atlas of the United States Based on the Results of the Ninth Census 1870*. New York: Julius Bien, 1874, Plate 44.

In carrying out his mandate, Walker stayed relatively close to his largely cartographic mission, but still found room to introduce novel graphic forms or redesign older ones to portray the American statistical landscape.

Particularly noteworthy is the idea to show two frequency distributions back-to-back, now called generally a *bilateral histogram*, or an *age pyramid* when the classification is based on age. Figure 7.13 shows one particularly complex example that indicates the level of specificity the atlases attempted.

Each bilateral histogram compares the number of deaths for males and females across months of the year; the sex that dominates is shaded. The top portion shows these for all the US states; these are arranged alphabetically, except for the last row, which contains small, mostly Western states. The middle portion shows these classified by cause of death. In the detail shown at the left, it can be seen most clearly that the shapes of these histograms vary considerably across diseases, and that some take their greatest toll on life in the winter months. The bottom portion is composed as a set of line graphs, classifying deaths vertically according to nationality (i.e., native white, colored, foreign-born) and horizontally by age, groups of diseases, specific diseases, and childhood diseases.

Another nice example (Plate 14) from this volume for the ninth census uses mosaic diagrams or treemaps to show the relative sizes of the state populations (by total area) and the breakdown of residents as foreign-born, colored, or white (vertical divisions). The last two groups are subdivided according to whether they were born inside or outside that state, with a total bar for inside / outside added at the right.[24] Other plates in this atlas (e.g., 31, 32) used similar graphic forms to show breakdowns of population by church affiliation, occupation, school attendance, and so forth, but we view these as less successful in achieving their presentation goals.

A dominant message is conveyed by the size of the diagrams for the states: New York, Pennsylvania, and Ohio are the most populous, and these states have similar proportions of foreign-born, colored, and white inhabitants. The subdivisions give a drill-down view of the details. Missouri, shown in the blow-up portion at the left of Plate 14, had a relatively larger proportion of white inhabitants born inside the state. Georgia, Virginia, and other southern states, of course, had larger proportions of colored inhabitants.

Following each of the subsequent censuses for 1880 and 1890, statistical atlases were produced with more numerous and varied graphic illustrations under the direction of Henry Gannett. These can be considered "the high-water mark of census atlases in their breadth of coverage, innovation, and excellence of graphic and cartographic expression."[25] The volume for the tenth census of 1880 contained nearly 400 thematic maps and statistical diagrams composed in 151 plates grouped in the categories of physical geography, political history, progress of the nation, population, mortality, education, religion, occupations, finance and commerce, agriculture, and so forth. The volume for the eleventh census of 1890 (Gannett, 1898) was similarly impressive and contained 126 plates.

However, the age of enthusiasm for graphics was drawing to a close. The French *Albums de Statistique Graphique* were discontinued in 1897 because of the high cost of production. Lovely statistical atlases appeared in Switzerland in conjunction with public expositions in Geneva and Berne in 1896 and 1914, respectively,[26] but never again. The final two US Census atlases, issued after the 1910 and 1920 censuses, "were both routinized productions, largely devoid of color and graphic imagination."[27] After the First World War, a few more graphical statistical atlases were published in emerging countries (e.g., Latvia, Estonia, Romania, Bulgaria) as a concrete symbol of national affirmation and a step in the construction of national identity. The Golden Age of Graphics, however, had come to a close.

The Modern Dark Ages

We defined a golden age is a period of high accomplishment surrounded on both sides by relatively lower levels: a mountain or a plateau. This is true for the Golden Age of Graphics. You can see this in the dip in graphical innovations into the 1950s shown in Figure 7.1. If the last half of the nineteenth century can be called the Golden Age of Statistical Graphics, the first half of the twentieth century can equally be called the "Modern Dark Ages" of data visualization.[28] What happened?

As mentioned earlier, the costs associated with government-sponsored statistical albums eventually outweighed the enthusiasm of those who paid the bills. But more importantly, a new zeitgeist began to appear, which would turn the attention and enthusiasm of both theoretical and applied statisticians away from graphic displays, back to numbers and tables, with a rise of quantification that would supplant visualization. Modern statistical methods had arrived.

It is somewhat ironic that this change of view reflects a form of intellectual parricide. The statistical theory that had started with games of chance and the calculus of astronomical observations developed into the first ideas of statistical models, starting with correlation and regression, due to Galton,

Pearson, and others, and this development was aided greatly by the birth of visualization methods and dependent on visual thinking.

Yet, by 1908, W. S. Gosset (publishing under the pseudonym Student) developed the t-test, allowing researchers to determine whether two groups of numbers (yields of wheat grown with or without a fertilizer) differed "significantly" in their average value. All that was needed was a single number (a probability or p-value) to decide, or so it seemed.

Between 1918 and 1925, R. A. Fisher elaborated the ideas of analysis of variance and experimental design, among his many inventions, turning numerical statistical methods into an entire enterprise capable of delivering exact conclusions from experiments testing multiple causes (fertilizer type and concentration, pesticide application, watering levels) all together. Numbers, parameter estimates—particularly those with standard errors—came to be viewed as precise. Pictures of data became considered—well, just pictures: pretty or evocative perhaps, but incapable of stating a "fact" to three or more decimals places At least it began to seem this way to many statisticians and practitioners.[29]

However, while there were few new graphical innovations in this period to count as milestones in this history, something else of importance happened: data graphics became popularized and entered the main stream.[30] This change in visual explanation did not quite have the popular impact of Einstein's theory of relativity ("it's all relative" became a common phrase to explain mundane observations with different viewpoints). Nevertheless, between 1901 and about 1925, a spate of popular books and textbooks on graphical methods began to appear. Quite soon, college courses on graphical methods were developed, and in the same period statistical charts, mostly mundane, began to decorate business and government reports.

Second, as we described in Chapter 6, graphical methods proved crucial in a number of new insights, discoveries, and theories in astronomy, physics, biology, and other natural sciences, many of these using the format of a scatterplot. In general, the *use* of graphical methods in the natural sciences continued throughout this period, though relatively little new ground was broken.

Interest in graphical methods arose again in the period from about 1950 to 1975 (labeled "Re-birth" in Figure 7.1) with a sharp rise in new innovations and a new respect for the power of a graph to show the unexpected or at least give greater nuance and insight into increasingly complex data. Over time, the tables were turned, at least slightly, on statistical models and single-number summaries in favor of visualization methods to expose the data to greater scrutiny. In 1962, John Tukey (1962) asked "Is it not time to seek out novelty in data analysis?" (p. 3) and began to answer this with a new paradigm of Exploratory Data Analysis focused on graphical methods. The modern period of data graphics was beginning, and it would take visualization to higher dimensions of display and data.

8

Escaping Flatland

In his much loved 1884 book *Flatland: A Romance of Many Dimensions*, Edwin Abbot described the mental sensation of taking a geometrical idea to one more dimension through movement:

> *In One Dimension, did not a moving Point produce a Line with two terminal points?*
> *In Two Dimensions, did not a moving Line produce a Square with four terminal points?*
> *In Three Dimensions, did not a moving Square produce—did not the eyes of mine behold it—that blessed being, a Cube, with eight terminal points?*

Thus, the inhabitants of *Flatland* had to contemplate a three-dimensional world that might exist outside the confines of the purely two-dimensional world of their perception and experience

In *Flatland*, a moving square could produce a blessed being, something that could only be "seen" in visual imagination, but nonetheless it provided an opening into a new world. Escaping flatland was yet another essential step in the development of visual thinking.

Indeed, but more abstractly, in both statistics and in data visualization, much of the progress can be thought of as an expansion in the number of dimensions contemplated,

$$1D \rightarrow 2D \rightarrow 3D \approx nD,$$

representing univariate, bivariate, and then multivariate problems.[1] The essential insight, initially in statistics, was that once you had solved some three-dimensional problem, a solution for the general, multidimensional case was not far behind.

Yet, although all of us live in a 3D world, and some people can think about four dimensions (just add time!) or even more, graphical methods had long been confined to a 2D surface: the flatland of a clay tablet, a papyrus scroll, a piece of paper or a computer screen. This chapter describes how data graphics escaped flatland.

In data visualization, the 1D stage was set with van Langren's graph (see Figure 2.1) of the estimates of longitude distance from Toledo to Rome. That was all he needed to make the argument to King Philip IV that all previous calculations had extravagant errors and the problem of longitude deserved a more accurate solution.

The next step, which we characterize as "1.5 D" (more than 1D but less than 2D), occurred in graphs of some variable along a conventional axis like time or distance. An early example is Phillip Buache's graphic depiction of the low and high tide levels of the river Seine (see Figure 5.8). Playfair (Chapter 5) took this to the next level, plotting multiple time series as line graphs (Figure 5.3), with a desire to show imports and exports, yet highlight the balance of trade. His 1821 chart (Plate 10) of wages, prices of wheat, and the reigns of monarchs is a tour-de-force example of the display of multiple time series, using different graphic forms (lines, bars, segments) to create a pleasing graphic. But the horizontal axis was still bound to time.

The idea of a fully two-dimensional plot—a scatterplot—of a dependent variable plotted on the vertical axis, against a predictor variable on the horizontal axis was the next step in this development, initiated by J. W. F. Herschel but fleshed out by Galton (Chapter 6).

Shortly after statistical methods developed from Galton's bivariate correlation ellipse (shown in Figure 6.16), Karl Pearson in 1901 sought to solve a more general problem, the first truly multivariate one in the history of statistics: Given a set of points in 2D, 3D, or higher-dimensional space, how can we find the line, plane or hyperplane of "closest fit" to the points?

Pearson's solution to this problem was an application of the method of least squares, but he argued it geometrically and visually. Unfortunately, he was only able to show it in two dimensions. His readers, like the inhabitants of *Flatland*, had to imagine a world of 3, 4, 5, and more dimensions in which this view could arise.

Figure 8.1 is the first illustration in his paper, showing a collection of points, $P_1, P_2, \ldots, P_n$. He argues that the variation of the collection of points

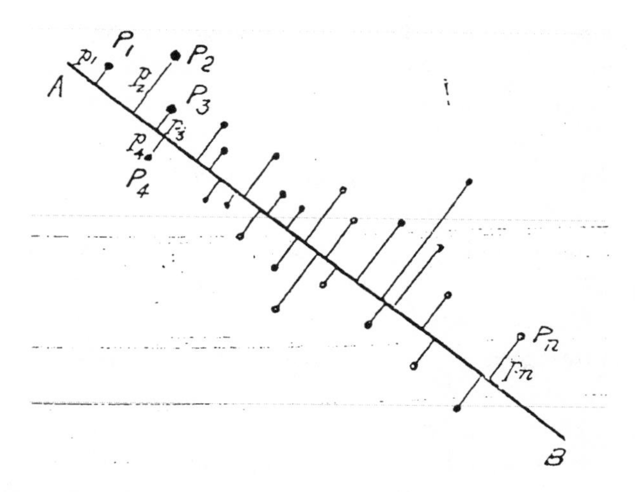

8.1 **Picturing a higher-dimensional plane in 2D:** Karl Pearson tries to show a solution to the problem of fitting a plane of closest fit. In 2D, the fitted plane appears as a line. *Source:* Karl Pearson, "LIII. On Lines and Planes of Closest Fit to Systems of Points in Space," *London, Edinburgh, and Dublin Philosophical Magazine and Journal of Science*, 2:11 (1901), 559–572, page 560.

is captured by the size and shape of the enclosing cloud or ellipsoid. The line or plane of best fit must therefore be perpendicular to the direction of deviations of the points, thus making the average squared length of the segments $p_1, p_2, \ldots, p_n$ as small as possible. Along the way, a number of problems spurred the development of three-dimensional visualizations.

Contour Maps

Maps start with a two-dimensional surface defined by latitude and longitude. After geographic features such as rivers, cities, and towns had been inscribed, it was natural for cartographers to want to show features of elevation, and landforms such as mountains and plateaus, in what came to be called *topographic maps*. This idea was a natural initial impetus for 3D thinking and visual depiction.

The first large-scale topographic map of an entire country was the *Carte géométrique de la France*, by the French astronomer and surveyor César-Françis Cassini de Thury [1714–1784],[2] completed in 1789. But well before these precise determinations of altitude were made, map makers began to try to show topographical features using contour lines of equal elevation on their maps. These were useful for finding the way through a mountain range as well as for military defense.

Beyond wayfinding and route navigation, *thematic maps* use the features of geography to show something more: how some quantity of interest varies from place to place. Figure 3.3 by Balbi and Guerry is a nice example of the use of shaded (choropleth) maps of France to display the geographic distribution of crimes and compare this with the distribution of literacy. But this and similar maps treat geographic regions as discrete, and simply shade the entire area in relation to a variable of interest.

The language and symbolism of maps expanded to display more abstract quantitative phenomena that varied systematically over geographical space. This was technically a small step from topographic maps that showed elevation of terrain using either color shading or iso-curves (lines of equal magnitude), but the impact was profound in scientific investigation. It was essentially what Galton had done in mapping the contours of equal barometric pressure across Europe (see Plate 12).

This idea, of drawing level curves or contours on a map to show a data variable, began much earlier. Perhaps the first complete example[3] is the 1701 map by Edmund Halley, showing lines of equal magnetic declination *(isogons)* for the world, shown here in Figure 8.2. It was titled, in a style that tried to tell the whole story on the frontispiece, *The Description and Uses of a New, and Correct Sea-Chart of the Whole World, Shewing Variations of the Compass.*

The purpose of Halley's map was to contribute to solving the problem of determining longitude at sea, the same question that had occupied van Langren. As explorers ventured farther, they discovered that the compass did not always point to true geographic north; a 1° abnormality implied a huge difference over a range of hundreds of miles. This angular difference between geographic north and magnetic north is called *magnetic declination*, but worse—its amount varied around the globe. Halley's idea was that a map of the world showing lines of constant declination would allow mariners to correct their estimates of longitude.

But a nagging question, which has long confounded map historians, is how did he do it? How did he construct such an elaborate and detailed map, covering most of the world with a fine mesh of lines of equal declination (at one-degree intervals)? As it turns out, Halley's map can be regarded as a

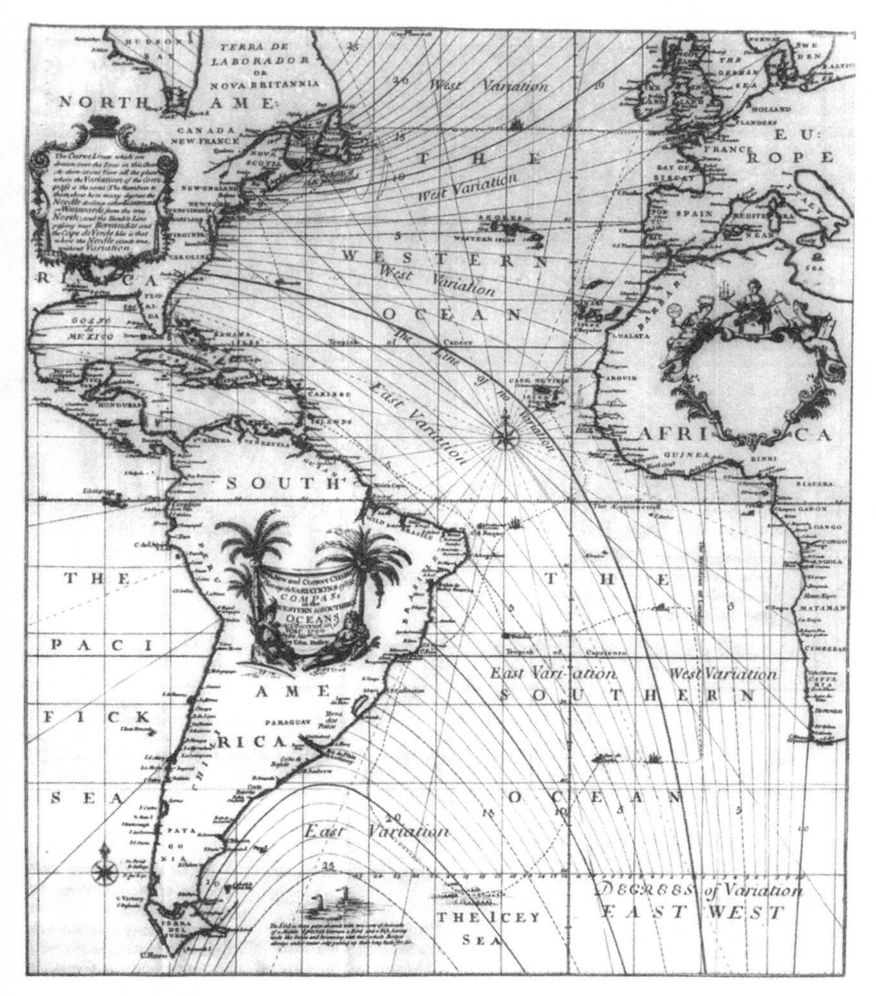

8.2 **Contour map of magnetic declination:** Edmund Halley drew lines of equal magnetic declination on a map, possibly the first contour map of a data-based variable. The figure shows the map for the Atlantic Ocean. The curve are the isogonal lines, with the degree of magnetic declination given as numbers along each. The thick line is the agonic line of no variation where the compass reading is true; the dashed line with ships shows the track of Halley's second voyage. *Source:* Edmond Halley, *A New and Correct Chart Shewing the Variations of the Compass in the Western & Southern Oceans as Observed in ye Year 1700 by his Maties Command*, 1701.

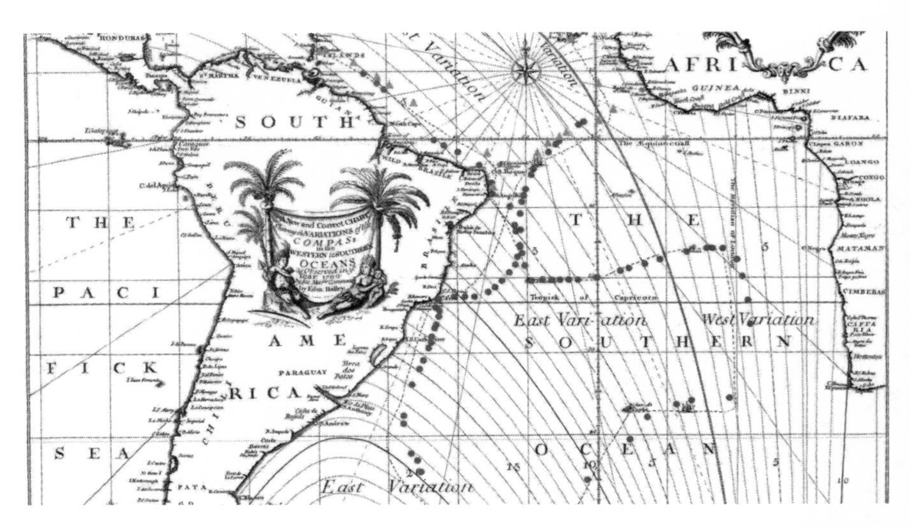

8.3 **Detail showing Halley's observations:** This figure shows the central portion of Halley's map with the locations of his observations. The triangles show the locations of observations from the first voyage; circles show those from the second voyage. *Source:* Detail from Lori L. Murray and David R. Bellhouse, "How Was Edmond Halley's Map of Magnetic Declination (1701) Constructed?," *Imago Mundi,* 69 (2017):1, 72–84, Plate 10.

monumental example of data interpolation and imputation. He collected his data in two scientific expeditions between 1698 and 1700 as captain of the HMS *Paramour.*[4] Figure 8.3 shows the central portion of the Atlantic map, with triangles and circles at the geographic locations where Halley made his observations. It is immediately clear that the actual data were thin.

In 2017 Lori Murray and David Bellhouse[5] presented an impressive example of statistical historiography. They digitized Halley's map and attempted to reconstruct his method from his observations, using only mathematical tools available in his day. They concluded that Halley used averaging as a form of smoothing his numbers, which he paired with Newton's method of divided differences to fit polynomial curves through the data. For example, given three or four widely spaced observations with the same reading, say a declination of +5 degrees, he could fit a polynomial curve and interpolate all the locations on the map that should give the same reading. In effect, Halley relied implicitly on a lawful regularity of magnetic declination: whatever the cause of this phenomenon, he believed that it must vary smoothly over the world.

Another lovely example of contour maps and smoothing is Galton's 1881 isochronic chart of travel time shown in Plate 15. The goal was to show how long it would take for a traveler from London to reach any destination in the world.[6] For his data, Galton used many sources, but most ingenious was an experiment in which he sent dated letters to many of his wide-ranging correspondents and asked that they reply indicating the date that the letter arrived. He considered the transport of a letter comparable to that of a traveler, and there was probably some degree of approximation or simple guessing in filling in the boundaries of the regions. Nevertheless, his chart answers a useful question, and provides another example of the methods Galton developed to visualize a third dimension of variation across the plane of a map.

Today, the problem of finding the contours of level curves of some response variable, z, over a two-dimensional (x, y) plane has relatively straightforward computational solutions, but that was certainly not so in the period we are discussing. Galton and others used seat-of-the-pants methods relying on averaging and often visual smoothing. Modern computer methods generally require measurements of z on a two-dimensional grid of equally spaced (x, y) values, and use interpolation in these values to find the locations of equal z values.[7] A solution to this problem was required in order to study empirical relations among three variables and perhaps establish mathematical laws.

In the corps of French engineers working at the French school of bridges and roads, which is known as the École Nationale des Ponts et Chaussées (ENPC), a workable general method for calculating level curves (*courbes de niveau*) was developed by Léon Lalanne [1811–1892] and extended by Louis-Léger Vauthier [1815–1901]. Like Minard, Lalanne worked primarily as a civil engineer on the construction of railways, working in Spain and Switzerland as well as in France; in his later career (also like Minard), he became increasingly occupied with the application of graphical methods to mathematical problems. In 1876, he became director of the ENPC and in 1878–1880 he published versions of a short book, *Méthodes Graphiques . . . à Trois Variables*.[8] The English title is self-descriptive: *Graphical methods for the expression of three-variable empirical or mathematical laws with applications to the art of the engineer and the resolution of numerical equations of a certain degree.*

Figure 8.4 shows one early example of his method, published in 1845.[9] Lalanne regards the data as a bivariate table that represents a 3D surface giving soil temperature in relation to date and hour of the day. The maximum temperature occurs in early July, at about 3:00 PM, and, around this time, the roughly elliptical contours vary more throughout a day than they do over dates.

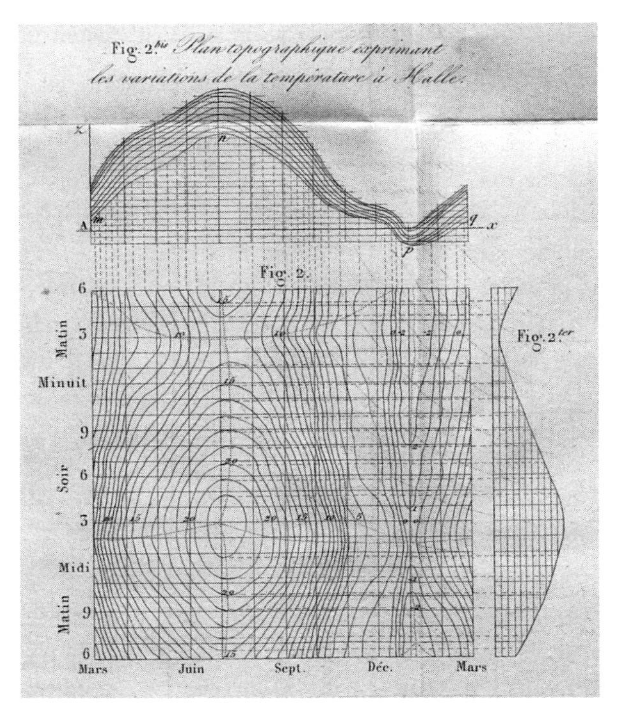

8.4 **Contour map of a bivariate table:** The graph shows the level curves of recordings of soil temperature measured over time, for months of one year (horizontal axis) by hours of the day. The maximum temperature occurs in early July, around 3:00 PM. *Source:* L. F. Kaemtz, *Cours complet de météorologie*. Paris: Paulin, 1845, Appendix figure 2.

At the top, Lalanne shows individual plots of temperature by date, with separate curves for different hours of the day, as if these were side views of the 3D surface projected on the wall of temperature versus date. On the right side, he shows one graph of temperature against time of day.

Louis-Léger Vauthier [1815–1901] also contributed to this method in a number of applications. In Figure 8.5 he tackled the problem of showing population density on a map, here for Paris in 1874. He did not make use of shading or other adornment here, except to show a few landmarks iconically (Notre Dame, Invalides). The various curves of population density are analogous to those of elevation on a topographic map and are labeled numerically (from 200 to 1,200). One can see the great concentration of the population on the right bank of the Seine, and another concentration on the left

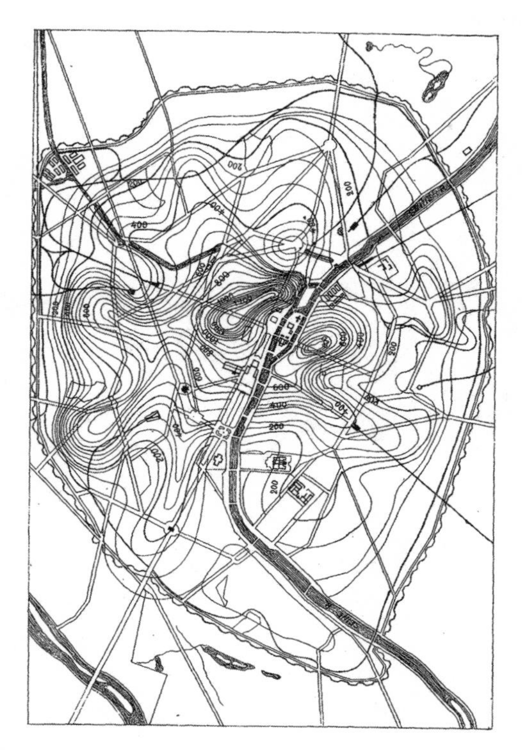

8.5 **Population density of Paris:** Louis-Léger Vauthier showed the population density of Paris by many contour levels representing densities of 200 to 1,200 people per unit area. *Source:* Louis-Léger Vauthier, 1874.

bank. This figure is remarkable for the number of levels of population density shown.

Three-Dimensional Plots

Contour maps and contour plots were certainly useful, but they were still images on a two-dimensional surface, using shading or level curves to show a third dimension. There is a huge difference between trying to navigate a driving route or a hike from a 2D map that shows elevation with isolines versus a 3D relief map that shows elevation in context, using perspective, realistic lighting ("raytracing"), color ("terrain colors"), texture mapping, and other techniques to generate beautiful and more useful 3D topographic maps.[10]

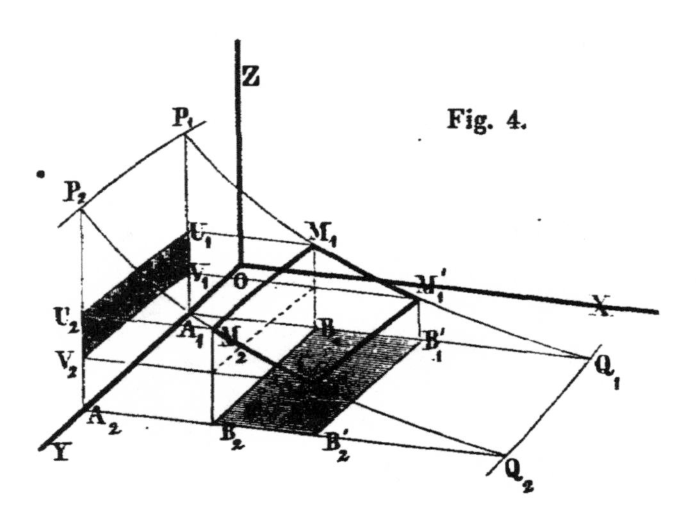

8.6 **Axonometric projection of a 3D surface:** The labeled points and connecting lines are meant to illustrate how surfaces and lines appear when projected onto the planes formed by the coordinate axes. *Source:* Gustav Zeuner, *Abhandlungen aus der mathematischen statistik.* Leipzig: Verlag von Arthur Felix, 1869, fig. 4.

The technique of rendering 3D views in depth and perspective on a flat surface was known to artists for centuries, but early landscapes lacked realism. The first exemplar to get perspective approximately right was the painting *View of the Arno Valley* by Leonardo da Vinci in 1473, his first known drawing; but that is just an artist's view. For data graphics, the precise technical details of drawing a 3D surface of a response variable z over a plane defined by (x, y) coordinates did not develop until the late 1800s. By 1869, in the course of work on thermodynamics, the German physicist Gustav Zeuner [1828–1907] worked out the mathematics of what has come to be called the axonometric projection: a way of drawing a 3D coordinate system so that the coordinate axes looked to be at right angles, and parallel slices or curves had the proper appearance. Zeuner took Descartes to 3D.

An example is shown in Figure 8.6. The coordinate axes, X, Y, Z are shown with the origin in the back. Two parallel curves are drawn, and the goal of this diagram is to explain how the rectangular region can be seen in terms of its projected shadows (shaded) as rectangles on the bottom and left planes.

The first known use of a 3D data graphic using these ideas was designed by Luigi Perozzo [1856–1916], an Italian mathematician, statistician, and, ultimately, a hero of demography, largely for this contribution to the study of the distribution of age over time.

A graphic innovation on this topic appeared in the U.S. Census atlas of 1870, where Francis Walker pioneered the idea of an "age-sex pyramid" showing the age distribution of the population by sex. It was called a pyramid because it compared the populations of men and women in back-to-back histograms by age, in a way that resembled a pyramid. In a number of plates, these data were broken down by state and other factors, in such a way that insurance agencies could begin to set age-, sex-, and region-specific rates for an annuity or life insurance policy. To demographers, this method gave a way to characterize fertility, life expectancy, and other questions regarding population variation. But these were still 2D graphs.

Perozzo took this a step further, into a third dimension, by considering how the age distribution of a population might change over time. He was able to get reasonably complete data from Sweden, for the years 1750–1875.[11] His goal was to go beyond Walker and show how life expectancy and the age distribution had changed over a period of 125 years. Figure 8.7 shows the initial 1880 version of Perozzo's stereogram, which was probably the first three-dimensional representation of data.

In this figure, age is drawn coming out of the plane of the figure, with the youngest age group (0–5) at the back, and the oldest (>80) at the front. Calendar years of the census are shown from left to right (1750–1875) and height in the figure is the size of the population at a given age. The left-to-right line at the back of the figure ("*Linea della Nascite*") gives the number of births.

The census counts for a given year, say 1750, are shown by the lines that go down and to the left, parallel to the age axis. Each slice in depth here is what Walker would have shown for a given population in a given year.

But Perozzo did something more. The addition of two other "time" axes allows different ways to view, think about, and understand age distributions beyond mere calendar time. A cohort is the group of people born in a given year. What are the ages to which they live? The distribution for a given cohort is shown by the thicker black line that runs down and to the right. What about the change in the number of people who live to a given age, over time? This is shown by the lines for a given age that go from left to right.

In this figure, one can see the general increase in life expectancy, either along the year lines or by cohort. Over time, the numbers of younger people (ages 0–5) increased dramatically, while the elderly (75–80) showed only

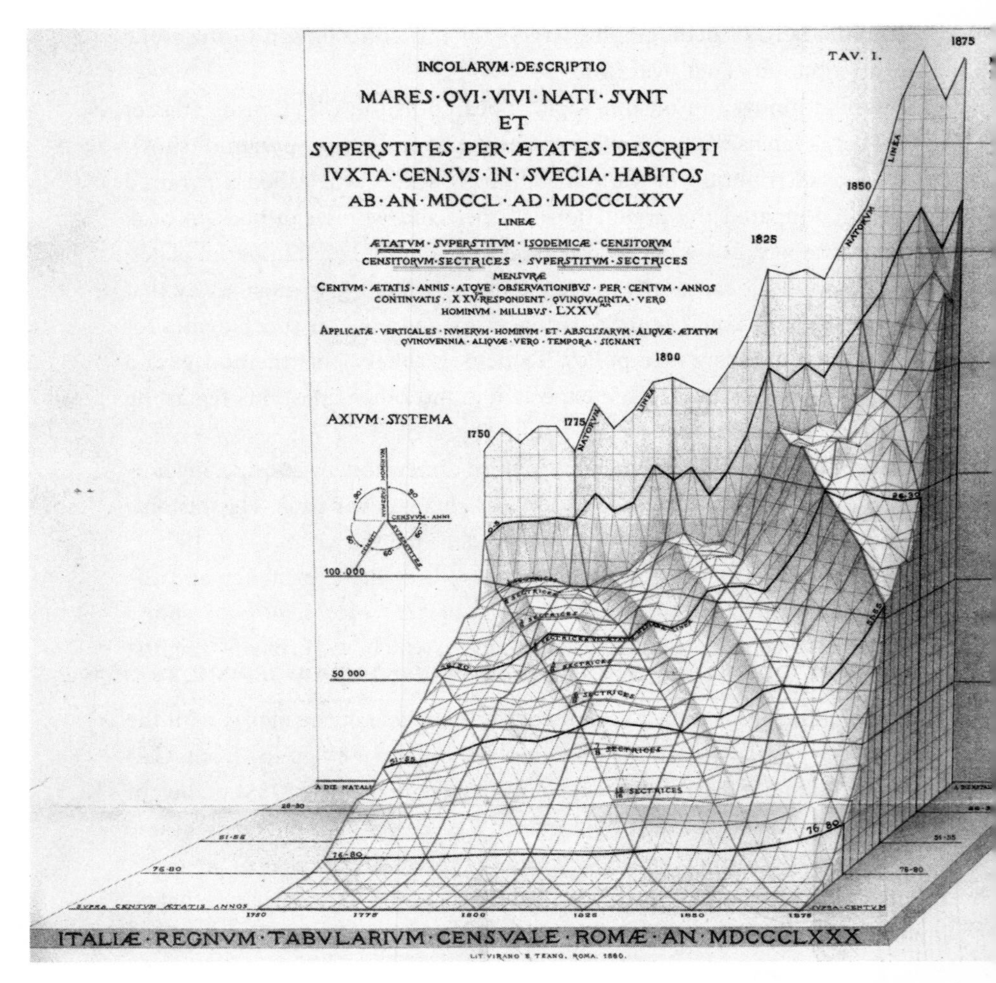

8.7 **3D population pyramid:** Luigi Perozzo showed the age distributions of the population of Sweden from 1750 to 1875 as a three-dimensional surface. Census years go from left to right, age is shown front (old) to back (young), and the height of the surface represents the count of people of that age. *Source:* Luigi Perozzo, "Stereogrammi Demografici – Seconda memoria dell'Ingegnere Luigi Perozzo. (Tav. V)," *Annali di Statistica*, Serie 2, Vol. 22, (Ministero d'Agricoltura, Industria e Commercio, Direzione di Statistica), 1881, pp. 1–20.

8.8 **3D statistical sculpture:** Perozzo created this 3D model of the population data as a tangible object, perhaps the first statistical sculpture. *Source:* Centre Pompidou.

modest increases to 1850 but then began to rise slightly to 1875. Other age groups were intermediate in these trends. For the cohorts born in a given year, the trace lines appear roughly parallel, except among the youngest age groups. A peculiar feature in this graph is the dip in younger people from about 1850–1870—it looks like someone chipped off a large chunk of the mountain. What could be the reason? Several demographer friends speculated: perhaps a war or an outbreak of disease? It turns out that the probable explanation is less dramatic: Sweden had large outmigration in that period.[12]

As did authors of other novel graphical methods, Perozzo was careful to explain the method for construction of his 3D chart. In the small legend titled "Sistema d'Assi" (system of axes) he indicates what the three axes are supposed to represent. In a second article in 1881, Perozzo[13] described the details of Zeuner's axiometric projection and compared it to other possible representations. He also presented a more elaborate version of his 3D population chart.

Perozzo recognized the value of a physical 3D sculpture. He created in papier-mâché or plaster a portion of his chart, as a tangible object, with lines showing the traces of age cohorts and census years (Figure 8.8). This probably would have remained unknown, except that André Breton, a French writer and exponent of the surrealist movement in art, found this item in a flea-market in the Saint-Ouen district of Paris. He considered this to be an essentially surrealist object because it connected the data of statisticians with ideas of artists who wanted to portray commonplace events in a new way.[14]

Going Forward

Perozzo's diagrams are still to be admired, partly for their execution, but also because they introduced a new way to see several aspects of "time" simultaneously: age, period, and cohort. In demography, these relations are now more commonly studied with a Lexis diagram,[15] a 2D plot of age versus time period, with cohorts represented as diagonal lines.

Three-dimensional plots that deserve to be shown in 3D are not very common in the scientific literature today, perhaps because it is difficult to render them as well as Perozzo did by hand. However, other important tools available now can make such data more easily understood: dynamic computer methods can use motion (change over time) to rotate a 3D image so that it can be viewed from various perspectives; methods of interactive graphics allow the viewer to control the view they see. This is the topic of the next chapter.

9

Visualizing Time and Space

The three decades from 1950 to 1980 were a period of active growth in the development and use of increasingly realistic data visualization. One thread concerned statistical and computational: dimension-reduction methods for representing high-D data in a low-D space that could be plotted, mostly in 2D.[1] Another thread in this period reflected new graphical methods, boosted by increasing computing power, which allowed graphic displays to become increasingly dynamic and interactive. Such displays were capable of showing changes over time with animation, thus changing the nature of a graph from a static image to one that a viewer can directly manipulate, zoom, or query. In these ways, the escape from Flatland continued as a wide range of important problems were illuminated by new approaches to understanding data in higher dimensions.

Once again, these developments illustrate the interplay between advances in technology (computer display and software engineering) and scientific questions for which visualization methods held promise. Today, we see the impact of this in the work of data journalists who now routinely present the details behind important stories (the Brexit vote in the United Kingdom, climate change, COVID-19, etc.) in high-impact online, interactive graphic applications. This chapter traces the origins of these ideas and some of the scientific questions that prompted this evolution of visualizing motion, time and space.

The Laws of Motion

There is, in nature, perhaps nothing older than motion, concerning which the books written by philosophers are neither few nor small; nevertheless I have discovered by experiment some properties of it which are worth knowing and which have not hitherto been either observed or demonstrated.

—GALILEO, *Dialogues and Mathematical Demonstrations Concerning Two New Sciences* (1638)

Galileo's seventeenth-century observations foreshadow the origins of the cinema, computer-animated films, and—most relevant to this narrative—dynamic data graphics. The popularity of modern dynamic data displays can be traced to scientific questions about the nature of human and animal motion. As technology developed, the study of motion and its visualization branched out from a pleasurable pastime to the huge industries of Hollywood and Netflix while also having important scientific applications in aerodynamics (the wind tunnel), medical imaging (blood flow in the heart and brain), and ecology (migratory patterns of animal species) among others

Aristotle's *De Motu Animalium* (*The Movement of Animals*) was the first book setting out the principles of animal locomotion. Nichole Oresme's 1360 "pipes" diagram (see Figure 2.2) was intended to show some possible mathematical relations between time and distance traveled. Around 1517, Leonardo da Vinci drew detailed anatomical studies of moving cats, horses, and dragons; Galileo later conducted experiments on motion and gravity between 1633 and 1642. However, the modern interest in these questions arose in the late 1800s when new technologies for recording could provide new insights.

To a physicist, motion is nothing more interesting than a change in position over time. It can be reduced to simple, but elegant, equations giving velocity (the first derivative) and acceleration (the second derivative). A velocity, v, of a horse galloping at 45 mph can be reduced to the equation $v = dx/dt = 45$. The acceleration, g, due to gravity on Earth can very nearly be reduced to a constant,[2] $g = 9.8m/s^2$, or $32ft/s^2$.

But to a ballet dancer, the art is in getting all the body parts to do those things in sync with a musical score to tell a wordless story of emotion[3] entirely through change in position over time. In data visualization, as in physics and ballet, motion is a manifestation of the relation between time and space, and so the recording and display of motion added time as a *fourth* dimension to the abstract world of data. We focus here on a few developments that led to the visual depiction, understanding, and explanation of time-changing phenomena.

The Horse in Motion

The modern scientific interest in visualizing motion can be traced to some simple yet perplexing questions of the late 1800s concerning the locomotion of the horse.

- How *exactly* do horses' feet move differently in a walk, a trot, a canter, and a gallop?
- What is the exact sequence of the four legs in each gait?
- How many feet are off the ground at any given time in each gait?
- Is there any moment, in each gait, when a horse is at least instantaneously suspended in air, with *all* four feet off the ground?

In the 1860s to 1870s, the last question was called "unsupported transit," and various writers weighed in to argue each side of the controversy.[4]

But there were no "data"; no information was available that was sufficiently precise to answer the question convincingly. The motion of a galloping horse was too rapid for either sight or sound to decipher, and even records of the positions of hooves on a specially prepared track could not be used to discern their exact pattern in time and space. In some ways, the armchair discussion on this topic resembled that of what to do about crime in France in the time of Guerry (Chapter 3), or the transmission of cholera in the time of Farr and Snow (Chapter 4).

The debate on horse locomotion was sufficiently intriguing to Leland Stanford (railway baron, governor of California, and a horse breeder) that he hired Eadweard Muybridge [1830–1904], a well-known photographer with a bent for technology, to try to answer the question by photographic means.

Cameras of that time recorded images on glass plates coated with a solution of silver nitrite; each new photograph required a new plate. Over some years, Muybridge devised a system using multiple cameras arranged in a line, with their shutters triggered by parallel strings stretched across the path of a running horse. Figure 9.1 shows one famous example of twelve successive frames of a horse, "Sally Gardner," galloping at a Palo Alto track (now the Stanford University campus). Frame 3 plainly shows all four hooves off the ground—settling the question. In other frames, only one foot is in contact with the earth, and the remaining frames provide a detailed account of the different motions of the front and back legs. A visual solution to the question of unsupported transit had been found. Legend has it that Stanford had bet $25,000 on the YES side and so was delighted with the outcome.

Nearly as surprising was the fact that that the all-feet-off-the-ground frame appeared among the frames when the horse's front and back legs were folded beneath the body, and showed a switch from pulling with the front legs (Frames 10, 11, 1, and 2) to pushing with the hind legs (Frames 4–6);

9.1 **The Horse in Motion:** Photographs of the horse "Sally Gardner" running at a gallop, June 19, 1878. *Source:* Library of Congress, Prints and Photographs Division, LC-DIG-ppmsca-06607.

contemporary artists had fancifully drawn images like frames 7–9, but with the forelegs extended forward and the hind legs backward to illustrate the possibility.

In a later reconstruction with higher-definition cameras,[5] Muybridge was finally able to make the horse's musculature visible to inspection, so one could see the powerful contractions of the muscles in the haunches as they propelled the horse forward at speed. A visual explanation of the galloping horse was finally available. But more importantly, the usefulness of such photographs as evidence in scientific questions had been established.

A Trick of the Eye

For scientific study, the background lines in the frames of Figure 9.1 were meant to allow some coarse quantification of the positions of the horse's legs over time. Although that information answered the question about the horse, it was far too imprecise for any finer analysis. Moreover, viewing motion in separate frames was a poor substitute for actually *seeing* a moving image.

Muybridge recognized the latter problem and invented a device he called a "zoopraxiscope," consisting of the separate images copied onto a circular disk

9.2 **Seeing motion:** Zoopraxiscope disk showing thirteen images of a kicking horse. The frames at the bottom show the sequence of extension of the hind legs together with the support of the fore legs. *Source:* Wikimedia Commons.

(see Figure 9.2), which when spun gave the illusion of motion, and can be considered the first movie projector.[6]

Muybridge, who was already famous on the West Coast for his breathtaking panoramic landscape photographs of Yosemite Valley, began to give lectures showing these early images of motion. It may be hard to imagine today, but his audiences were spell-bound to see ordinary motion slowed down and captured in this way. He explained the effect by saying that this was just a trick of the eye: the eye fools the brain into thinking that a series of connected pictures represents motion. This trick is called *persistence of vision*: an image falling on the retina "sticks" to it for about 1/10th of a second, giving an illusion of smooth motion from a sequence of separate images. This trick of the eye was novel because it provided a *new way* of seeing and understanding motion.

Today we know that the separate frames in Figure 9.1 are an *implicit* animation depicting change over time in a single, static 2D image. What one sees in the zoopraxiscope or an animated GIF is an *explicit* animation, images superposed over time, so the mind's eye can see this as motion. It is immediately clear how the horse gallops, but harder to see the precise moment that all feet are off the ground.

Muybridge's lectures in Philadelphia caught the attention of wealthy and influential benefactors of the University of Pennsylvania. He was wooed to relocate there with an offer of land (fittingly, on the grounds of the Veterinary Hospital) and a grant to construct an outdoor studio with a bank of electrically controlled cameras for the systematic study of animal locomotion. In the 1880s, Muybridge and his assistants made over 100,000 images of animals (from the nearby zoo) and human models. Nearly 800 plates comprising 20,000 images were published in *Animal Locomotion* (1887). This massive collection can be considered the modern beginning of the photographic study of animal motion; among other things, his work with human models later inspired Marcel Duchamp's *Nude Descending a Staircase* in 1913.

Lastly, in another notable achievement, at the 1893 World's Fair in Chicago, Muybridge gave a series of lectures on the topic "Science of Animal Locomotion," in a Zoopraxographical Hall built for that purpose. He used his zoopraxic device to show these moving images projected on a screen to an appreciative paying public, making this the first commercial movie theater. These tricks of the eye would later become important topics in perceptual psychology and vision science, but Muybridge's photographic studies planted new ideas about seeing and recording motion.

Étienne-Jules Marey and a Science of Visualizing Time and Motion

In France, at more or less the same time, Étienne-Jules Marey [1830–1904] had also begun to study how to make animal and human movement more apparent to graphic inspection and amenable to quantifiable scientific study. Whereas Muybridge had come to the visual study of motion through his artistic background and had developed some technology for recording and display, Marey came to this subject from a background in medicine and a burning desire to use the study of motion, not as a merely informative and entertaining technology, but as a gateway to scientific discovery. He was by turns an important physiologist, a prolific mechanical inventor, and a chief proponent of the idea for what he called "The Graphical Method," which provided a gateway to simultaneously record, quantify, and explain scientific phenomena.

Marey trained as a doctor at the faculty of medicine in Paris, but from the outset his interest was more in clinical research than in medical practice. He began this work in 1857 with studies of cardiology and circulation of the blood, using his engineering and mechanical skills to design new and improved instruments for measuring and recording blood pressure and flow. His improved design of the sphymograph, for example, used a set of levers and weights attached to a wrist cuff (similar to the blood pressure cuff used today) and an arm that magnified the pulse waves from pressure in the radial artery and recorded them on paper with an attached pen. Now, the internal state of blood pressure could be measured, tracked over time, and made visible.

In this, the beginnings of modern physiology, Marey was guided by a mechanistic belief that the human body was largely an animate machine governed by the laws of physics known from mechanics (force, mass, work), hydrology (e.g., to study blood flow), and an emerging understanding of thermodynamics (conservation of energy). More importantly, these connections could be used to discover analogous laws of physiology: "In the living organism we shall find those manifestations of force which are called heat, mechanical action, electricity, light, chemical action. Among the phenomena of life, those which are intelligible are precisely those of the physical or mechanical order."[7]

This appeal by analogy to the laws of physics is precisely of the same logical form used by Guerry (see Chapter 3) in seeking to discover law-like regularity in rates of crime and other social variables through the study of constancies and variation. "Data" was crucial for both, but for Marey, this meant first inventing new technology for the precise recording of change over time, giving motion—and then some means to study that from graphical inscriptions.

La Méthode Graphique

To achieve his goals, Marey devised and improved many instruments for recording motion, most often using one or more pens tracing paths of change on a rotating cylinder or a flat sheet. Among other topics, he studied the functions of voluntary muscles with a myograph to record the intensity of muscle contraction, developed a "recording shoe" to study change in pressure from the foot in walking, and even constructed a leather recording bracelet for a horse to inscribe the details of its gait.

Around 1870, he began to study aerial locomotion of flying insects and birds, using ingenious devices such as that shown in Figure 9.3. He sought to answer some of the fundamental questions of flight, including: what is the frequency of movements of the wings of various species of insects and

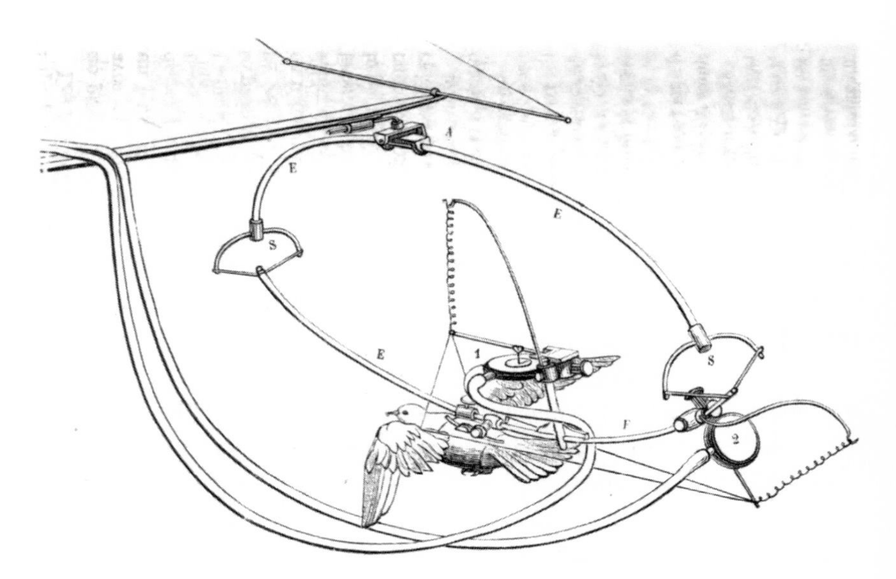

9.3 **Recording flight:** A harness, designed to register the trajectory, force and speed of a bird's wing in flight. *Source:* Étienne-Jules Marey, *La machine animale, locomotion terestre et aérienne*. Paris: Librairie Germer Baillière, 1873, fig. 104.

birds; what is the sequence of positions the wing takes in each phase of periodic movement; and by what mechanism does the wing, using the air as its counterforce against the body as a fulcrum, produce lift and forward motion?

Apart from purely scientific appeal, these questions were also of direct interest to aspiring aeronauts thinking of constructing flying machines. But the larger questions for Marey concerned how to connect anatomy (skeleton, joints, and muscles) to physiology and how to make these processes both visible and quantitative.

Marey constructed many mechanical models of artificial insects and birds as studies of the mechanisms of motion of real flying creatures, and he devised methods to record the motion of the wings of the common housefly, bees, wasps, and other insects by the rubbing of a point on its wing against a rotating smoked cylinder. Among other things, he calculated the number of beats per second: common housefly: 330; bee: 190; wasp: 110; and down to the cabbage butterfly at 9.

An 1869 report in *Scientific American*,[8] of an early study published in the *Comptes Rendus*, extolled the virtues of the graphic method compared

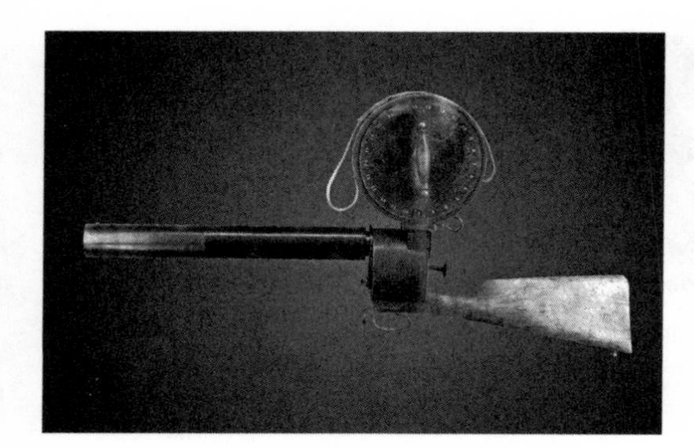

9.4 **Photographic gun:** One of Marey's photographic guns. The cylinder at the top holds the film roll, recording twenty-four images over a two-second period. *Source:* Wikimedia Commons.

with earlier and less reliable methods based on sound: "In the face of these discordances, the author sought for a mode of exhibiting, in an unmistakable manner, each of the beats of the wing of an insect, and the graphic method answers very well for determining their frequency" (p. 242).

Chronophotography

Recording traces of insect wings, pressure from a man's foot, or a horse in motion with pens transcribing that motion to paper did yield some new insights, but the technique of recording by moving pens was not fundamentally different from Robert Plot's recording of barometric pressure 200 years earlier (see Figure 1.4). Marey's application of graphic recording differed mainly in the widely inventive scope of his mechanical inventions to drive the pens and make hidden phenomena visible. These were certainly important, but they only captured some parts of dynamic phenomena. On the other hand, Muybridge's precisely sequenced photographs and his machines to display these as moving images were certainly useful and entertaining, but they lacked the capability to allow people to observe and study changes in the complex positions of many separate parts over time in a simple and direct way.

In 1882, Marey perfected a method to record successive frames in the *same* picture with a "photographic gun" (see Figure 9.4) capable of recording twelve frames per second on a single strip of sensitized film in real time. This put all the images together in a time-motion panorama that could be more easily seen and studied at once.

9.5 **Starting a sprint:** Chronophotograph of a runner beginning a sprint. *Source:* Étienne-Jules Marey, Flight/WordPress.com.

This established the principle of multiple photographs, sequenced in time in a single image yet showing motion. Marey used this to record many images, of pelicans and other birds, horses under various conditions, and human athletes doing various tasks. But the hand-held gun was soon abandoned in favor of a steady camera with a timed shutter, recording multiple images on a single fixed photographic plate. A new technique in the study of motion and dynamic phenomena had arrived.

Figure 9.5 is just one of many hundreds of studies of human and animal motion that Marey carried out from 1882 to about 1888 in his *Station Physiologique* in the Bois de Boulogne in Paris with his assistant Georges Demenÿ (who was also a gymnast). It shows a runner (naked, to make his muscles visible) starting a sprint from a crouched position at the right. He takes about 1/2 second (seven frames) to reach an upright position. The successive frames alternate between power push from the hind leg to landing on the opposite leg. Other athletic studies show pole vaulting, fencing, and other sports. This was the beginning of the photographic analysis of sport, which is now widely used to help athletes get the next inch of height in a pole vault or shave another 0.01 second off their time for a 100-meter dash.

Falling Cats

Like the problem of unsupported transit of the horse, the question of how a falling cat almost always lands on its feet attracted the interest of some famous mathematical physicists, including James Clerk Maxwell and George Gabriel Stokes. It was suggested that cats had a "righting reflex," and the empirically minded Maxwell and others attempted to determine such things as the minimum height of fall for a cat to fail to right itself. In this time, cat defenestration

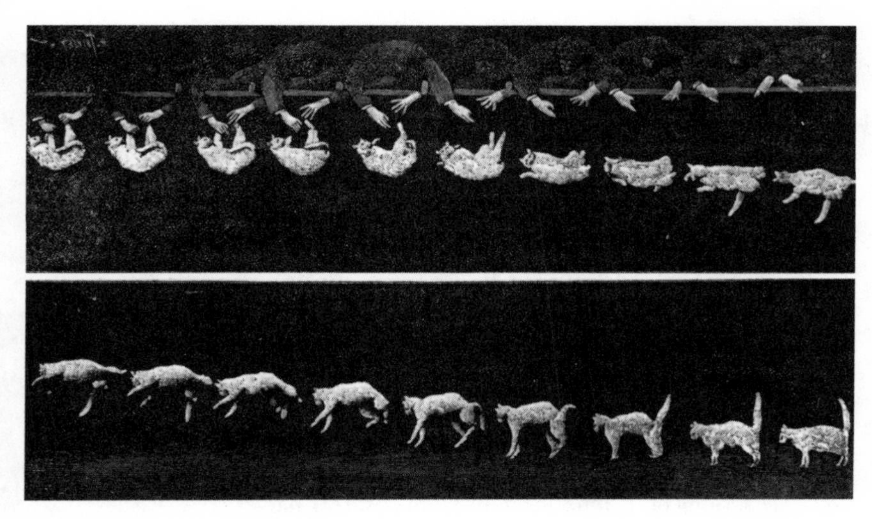

9.6 **Falling cat:** Successive images of a falling cat, captured at twelve frames per second on a chronophotographic device. The sequence starts at the upper left, and is continued at the left of the lower panel. *Source:* "Photographs of a Tumbling Cat," *Nature*, 51:1308 (1894), pp. 80–81, fig. 1.

was reputed to have caused death or injury to not a few felines in the name of science, and this practice was said to be banned at Trinity College, Cambridge.

But the exact mechanism of the cat's usual response and a physical explanation of it were unknown. To physicists, the problem was how a cat could right itself without violating the laws of physics. The principle of conservation of angular momentum meant that rotational change must remain constant unless acted upon by an external torque; so how could a cat rotate in 3D space in the absence of an external force?

To Marey, this was just another applied scientific problem he could tackle with chronophotography. In 1894, he carried out a series of empirical studies of falling cats (and later, of chickens and dogs). As shown in Figure 9.6, the cat was held by its feet and let go in that position. One gun was used to record the side view, shown here, and another recorded the end view. He also arranged these frames into a short film,[9] making this the first cat video.

The still chronophotograph in Figure 9.6 consists of nineteen images, corresponding to 1/12th-second intervals. The cat's vertical position is nearly the same in the first five of these while the assistant opens his hands, until the cat is finally in free fall. What is remarkable is that the cat's righting response is immediate and is nearly completed in frames 6–8.

Equally, however, detailed inspection provides a close account of the cat's plan for a happy landing. In the bottom row, as the cat approaches the ground, her legs are still stretched out, front and back (frames 11–13). In frames 14–16, she does two things: she extends her feet down for landing and arches her back to absorb the shock. Finally, in the last three frames she raises her tail, to achieve balance in her final position. Voilà! (Cat faces the camera, takes a bow, and exits stage left.)

In 1894, Marey[10] published his investigations of the motion of falling animals in the prestigious French *Proceedings of the Academy of Sciences* (*Comptes rendus de l'Academie des Sciences*). He was able to conclude that it was the inertia of the cat's own mass that enabled it to right itself, using separate actions of its muscles in its front and rear halves.

The important feature of this example is that others were able to "read" these images as well, and draw their own conclusions. An anonymous writer in *Nature*[11] stated that "the rotation of the fore and hind parts of the cat's body takes place at different stages. At first the twist is almost exclusively confined to the fore part, but when this amounts to about 180 degrees the rear part of the animal turns." He also concluded that, "The expression of offended dignity shown by the cat at the end of the first series indicates a want of interest in scientific investigation" (p. 80).

A solution in physics using a mathematical model of the "salient features of the motion of the falling cat," composed of two parts that could pivot in the middle, was finally proposed by Thomas Kane and M. P. Scher in 1969.[12] As a visual proof of their theory, they offered a computer-drawn animation of photos of a falling cat, overlaid with the shapes of the cat sections described by their model, as shown in Figure 9.7. The idea of computer-animated graphics as a means to think visually about complex, dynamic systems had begun.

Computer Graphics Animation

The next major advances in dynamic visualization took place in increasingly realistic computer-generated graphics. Some first steps, beginning in the 1970s, involved the development of techniques for 3D rendering of objects

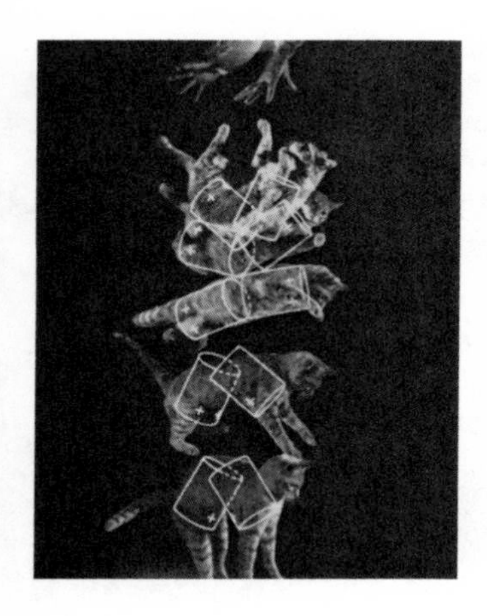

9.7 **Visual proof:** Computer-drawn images of rotation according to the mathematical model, overlaid on photographs of a falling cat. *Source:* T. R. Kane and M. P. Sher, "A Dynamical Explanation of the Falling Cat Phenomenon," *International Journal Solids Structures*, 5 (1969), pp. 663–670, fig. 6. Reprinted with permission from Elsevier.

with photo-realistic lighting, texture, and perspective. Such rendering begins with a 3D model of a physical object, using a collection of points in 3D space, connected by geometric entities such as lines, triangles, curved surfaces, and so forth. As a collection of data, an object can sometimes be calculated from formulas, and thence be manipulated by software to be rotated, zoomed, deformed, shaded, and so forth to move through space and contribute to telling a story.

For example, Figure 9.8 shows a 3D rendering of an apple, with colored stripes corresponding to colors of the rainbow. It is clearly not as realistic as a photograph of a real apple, but it is remarkable that the shape of a reasonable apple can calculated from simple equations for each of the three spatial coordinates of the points on the surface.[13] Once that has been done, color or texture can be mapped onto the surface, and then algorithms can be applied to simulate the paths of light rays from a given light source, producing a more realistic image. The apple can then be animated, causing it to move through space. If it is to appear in a scene, such as a cartoon or movie, the computer model is "rigged" by giving it control points that allow its shape to change systematically by interpolation as forces are applied. For example, one could animate the apple being thrown against a wall, compressing on impact, and bouncing off.

Most real objects cannot be defined by simple equations. However, just as a sculptor can create the basic form for a person, a chicken, or a cat from

9.8 **Parametric apple:** Computer-generated image of an apple using parametric equations for the 3D solid surface. Rendering of the surface uses methods to simulate illumination from a light source in front of the viewer. *Source:* © The Authors.

9.9 **3D graphics model of a cat:** Left: the form of the cat is defined by a wire frame consisting of points and connecting lines; right: the model of the cat has had fur applied and is then animated, twisting as it falls. *Source:* nonecg.com.

a molded wire mesh, so too a 3D graphic modeller can create a wire-frame image of any object by creating a data object consisting of (x, y, z) points on the surface and lines connecting those in a virtual wire mesh. For instance, Figure 9.9 shows a current high-quality wire-frame image of a cat in the left panel. The digital cat is just a large collection of (x, y, z) numbers and instructions for drawing the edges and faces in the image. In the right panel, the cat was made more realistic by applying color, a light source, and texture mapping for the fur. Then, the twisting motion of the cat while falling, as explained by Kane and Scher (Figure 9.7) can be animated by applying their equations of cat motion to the data of their rigged virtual cat.

These developments led to spectacular films blending real actors seamlessly with computer-generated imagery (from Steven Spielberg's 1993 *Jurassic Park* to James Cameron's 2006 *Avatar*). The scientific benefits of this technology are still active topics.

Animated Algorithms

Statistics and data visualization were later arrivals in the development and use of motion and animation. First, because the use of graphical display for these purposes depended on advances in software and computer power for data analysis. But even more so, visualization required specialized hardware and other technology for dynamic and interactive graphic display.

Playfair, Guerry, Minard, and Galton had all drawn their beautiful charts and maps by hand. Minard had shown changes over time in cotton exports (Figure 7.5), and Galton had depicted the complex relations of multivariate weather data over time and space (Plate 12). But something more automated was needed for modern problems, especially to visualize change over time when the objects of attention were numbers calculated from statistical models rather than horses or cats.

Until the mid-1960s, most data analysis used large "mainframe" computers located in a central computer center. Programs and data were read in from punched "IBM cards," and the output was printed on fanfold paper. Statistical graphs, of a sort that now look primitive, could be drawn by programs that instructed the computer to print the characters, line by line, to produce a bar chart, a scatterplot, or even a map; a good bit of ingenuity went into doing more, with such tricks as overstriking many characters on a printer to create shading patterns on a map. Computer-driven pen plotters soon made high-resolution static graphs and maps easier to produce.[14]

The big development in hardware for data graphics was the cathode ray tube (CRT), which had long been used in television, oscilloscopes, and radar. But when hooked to computers they could provide primitive graphical displays, as shown in Figure 9.10. These early computer displays from 1950–1965 used only black and white and had limited resolution, but for the first time they provided the means to produce animated computer graphics. Very quickly,

9.10 **Early examples of CRT displays:** CRT displays from 1950 to 1965. *Source:* (top) National Institute of Standards and Technology / Wikimedia Commons; (bottom) Todd Dailey / Wikimedia Commons / CC BY-SA 2.0.

pen-like input devices (see the bottom panel of Figure 9.10) were added, allowing a user to interact with the information shown on the display.[15]

The MDS Movie

As far as we are aware, the first use of this new technology in statistical graphics was a movie produced by Joseph B. Kruskal at AT&T Bell Laboratories in 1962 to illustrate the algorithm used in the statistical method of multidimensional scaling (MDS). MDS aims to represent the perceived similarities among a collection of objects (brands of automobiles, political candidates) by the distances among points in a space of some initially unknown number of dimensions (1D, 2D, 3D, . . .). Kruskal's method was iterative because there was no exact solution: it started with an arbitrary configuration of points, and then, one by one, moved them a little to improve the match between the order of the similarities and the order of the distances among the points in space.

The MDS method promised to provide a way to discover both the *number* of spatial dimensions required to adequately represent the data and the *nature* of the dimensions underlying human judgments or other data reflecting similarity. It soon became important in a number of scientific areas. In the social sciences, it was used to find the perceptual or cognitive spaces of colors, tastes, facial expressions, phonemes in speech recognition, semantic memory, and attitudes toward world nations using only ratings of *similarity* or measures of *confusion*. "Dimensions of the mind" could now be studied using a variety of simple tasks. In archaeology, it was later used to quantify the relationships among a set of digging sites based on the number of shared features in artifacts found in those sites. In chemistry, MDS began to be used to find the spatial structure of molecules, and the method served as an early basis for methods of graph layout, which is now widely used in network visualization.

Kruskal and his colleagues at Bell Labs wanted to illustrate the mathematical method of moving the points in space so that their distances reflected the similarities could actually work. In August 1962, they produced a three-minute movie[16] showing the steps taken in solving this problem as a visual demonstration of their algorithm. Most of the movie is like a primitive YouTube video with slide panels describing what it is about. The actual animation lasts only 20 seconds and was recorded by connecting a film recorder to a computer display, akin to Marey's photographic gun hooked to a computer.

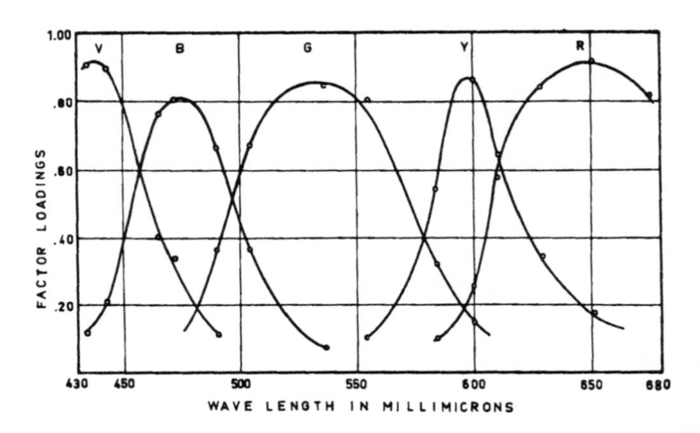

9.11 **Dimensions of color similarity:** Ekman's five-factor solution for the similarities among colors. *Source:* Gosta Ekman, "Dimensions of color vision," *Journal of Psychology*, 38:2 (1954), pp. 467–474, fig. 3. Reprinted with permission of Taylor & Francis Ltd.

As a demonstration, Kruskal chose a well-known problem from the psychophysics of color perception: Can human judgments about the similarity of color samples give reliable information about how colors are actually perceived by observers? If similarity judgments can be mapped into positions in a two-dimensional space, how does this relate to standard color theory?

A 1954 experiment by the Swedish researcher Gosta Ekman provided the necessary data.[17] Ekman chose fourteen colors varying only in wavelength from violet to dark red. If perception reflected physical properties, these colors should resemble the standard color circle.

Yet MDS had not yet been invented in 1954, and Ekman used what he knew, a method called factor analysis, to try to find the "dimensions of color vision." His result is shown in Figure 9.11, where he plotted the factor loadings of the colors against wavelength and drew smoothed curves in an attempt to show that his factors could be interpreted as violet (V), blue (B), green (G), yellow (Y), and red (R), in relation to wavelength.

This graph is not terribly wrong, in the same way that Ptolemy's depiction of the orbits of heavenly bodies around the Earth was not terribly wrong. It accounted, more or less, for the data, and Ekman's explanation of these sinusoid curves described some rough ideas about the sensitivity of color

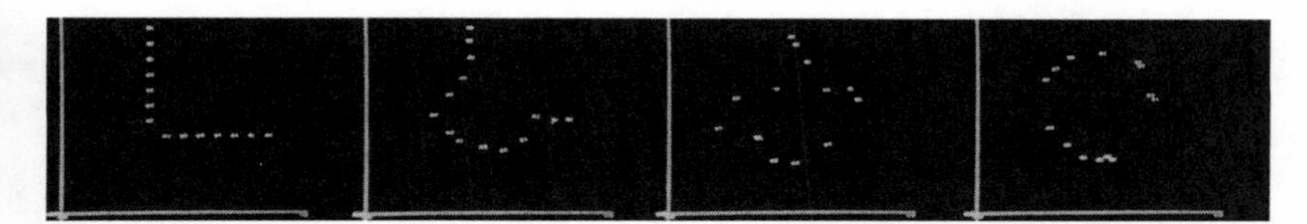

9.12 **The first statistical movie:** Four frames from the multidimensional scaling movie, the first animation of a statistical algorithm. Left: initial configuration for fourteen colors; right: final configuration, making the order of interpoint distances most closely agree with the order of the similarity judgments. *Source:* J. B. Kruskal, Multidimensional Scaling, AT&T Bell Laboratories, 1962.

receptors in the retina or the interpretation in the brain of responses to varying wavelengths of color. But it was the wrong explanation because it was bound to his method of seeing the results of his experiment from factor analysis. It also failed the test of parsimony: a 2D color circle was much simpler than five color factors.

Figure 9.12 shows four frames from the MDS video. At the left, the fourteen color samples are arbitrarily arranged in an L-shaped configuration. In each successive frame, the positions of the points in space change in such a way that the differences between the ordered similarities and the ordered distances in space are made progressively smaller. The 20-second animation part of the video comprises about sixty frames. In the final frame shown at the right of Figure 9.12, the configuration of points has converged to a stable solution, meaning that no further jiggling of the points makes it better.

As a "movie," the result is extremely primitive and thoroughly unexciting. However, for those involved at Bell Labs and those interested in the MDS method, the impact was dramatic. Galileo, had he been present, would likely have been prompted to exclaim once again "*Eppur si muove*!" (And yet it moves!).

More importantly, the result was in accord with color theory, which states that these color samples should reflect a simple 2D circular arrangement of hues in color space. Plate 16 shows the result of a modern reanalysis of Ekman's data using the same MDS algorithm. Compared with Ekman's factor analysis, which assumed that similarities reflected a true metric measurement scale (like a temperature), the nonmetric MDS result was simpler because it assumed only an ordinal relation between similarity and distance among points in the 2D space.[18]

Kruskal established the idea that even a primitive computer animation could shed light on a complex algorithm, otherwise presented in mathematical formulas. An even simpler case, but one requiring both graphic animation and a 3D representation would later arise in the study of algorithms for generating "random" numbers, which are used in what came to be called the Monte Carlo method.

The Monte Carlo Method

The Monte Carlo method is a technique for solving complicated problems by statistical sampling using random numbers. It was developed by Stan Ulam and John von Neumann in conjunction with calculations at Los Alamos in the high-energy physics of neutron chain reactions in nuclear fission devices, leading up to and following the development of thermonuclear atomic weapons.[19] Given some initial conditions and one hundred starting neutrons, how many would be available after, say, 10^{-6} second (1 microsecond)? The calculations involved the velocities of hundreds of neutrons scattering, being absorbed and colliding, and were far too complex for direct mathematical analysis.

Ulam had the brilliant idea that the history of each neutron could be traced by using random numbers to select the outcomes of the various interactions along the way, like balls in a pinball machine. Doing this a large number of times would give a statistical distribution of the states of a population of neutrons under given experimental conditions, whose properties gave reasonable answers to the questions posed.

In 1947, von Neumann, who was working on the development of the first programmable digital computers, sketched a program for the ENIAC (the Electronic Numerical Integrator And Computer) for generating random digits from 0 to 9 and then turning these into a distribution of a given form. He estimated that with this method, following one hundred primary neutrons through one hundred collisions each, the calculations would take about 5 hours. But, the virtue of doing this calculation by a computer program was that any other problem in this category could be solved by changing only a few numerical constants.

Monte Carlo simulation quickly found other uses in science and practice, including cryptography, the structure of crystals and chemical molecules, and mathematical evaluation of multiple integrals, thus turning intractable

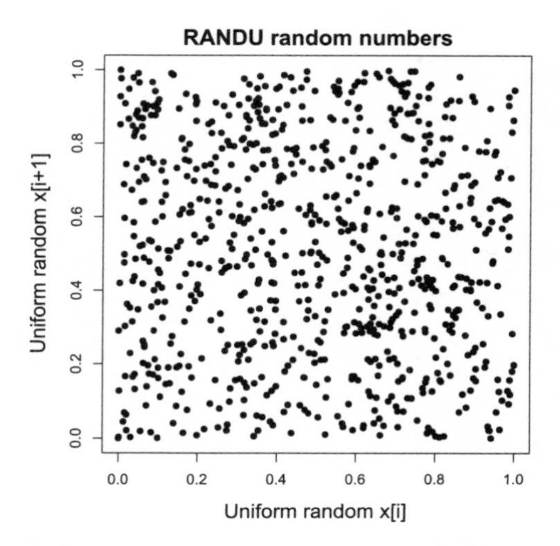

9.13 **RANDU:** Plot of successive pairs of 800 random numbers generated by the RANDU algorithm. Nothing unusual stands out: the points appear to be randomly and uniformly distributed in the unit square. *Source:* © The Authors.

problems into ones that could be solved by simulation using random numbers and tallying the results. But this technique hinged on having a method to generate an adequately random set of "random numbers" on a digital computer.[20]

RANDU

By the 1960s, IBM mainframe computers became widely available in universities and businesses. They included many utility programs, among which was a random number generator called "RANDU," based on the idea of sequential calculation, starting with some initial "seed" number, v_0, a simple formula to calculate the next number, v_{i+1}, in the series from the current one, v_i, would have all the properties of truly random numbers. RANDU used a formula that was very easy and fast to calculate in the binary arithmetic of computers. It was supposed to produce integers numbers that were uniformly distributed in the range $[1, 2^{31} - 1]$. In practice, these numbers were divided by 2^{31} to get random numbers in the interval $[0, 1]$.

Algorithmic calculation can give only *pseudo-random* numbers, but some methods come closer than others in behaving like quantities that are truly random, such as numbers obtained from tossing a very large number of dice.

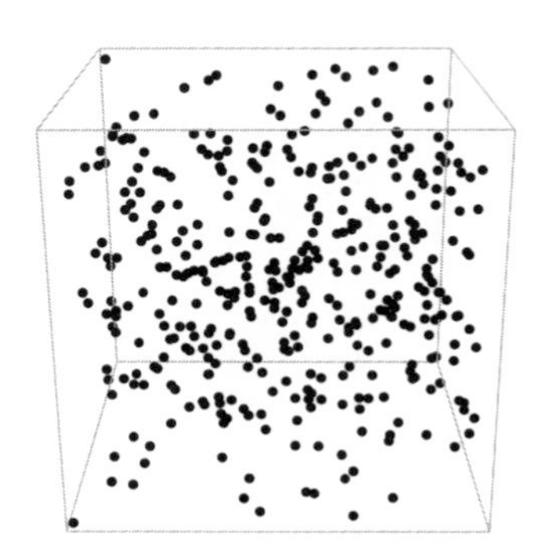

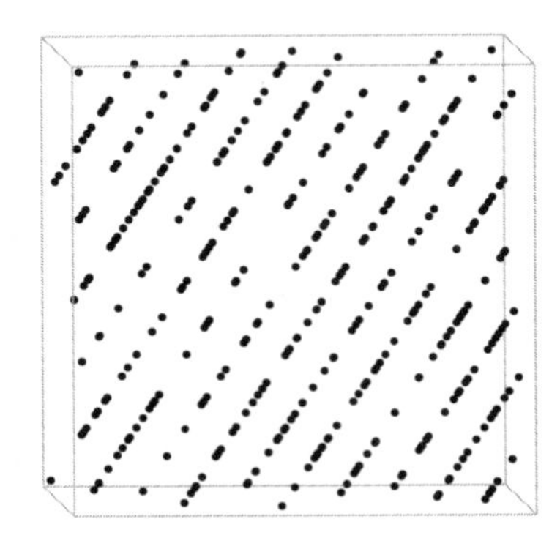

9.14 **RANDU in 3D:** Three-dimensional plots of successive triples of 400 sets of random numbers generated by RANDU. The left panel shows the positions of the points in a standard unit cube, with axes x[i], x[i+1], x[i+2]. The right figure shows the same plot rotated slightly. The points now appear on 15 parallel planes in 3-space, indicating their systematic (non-random) pattern. *Source:* © The Authors.

RANDU was put through a variety of tests, some statistical and some graphic. For example, plotting successive pairs of numbers (see Figure 9.13) should show nothing systematic in the distribution of points in the unit square. Random numbers should be totally featureless.

Yet RANDU is now considered the most ill-conceived random number generator ever designed. George Marsaglia discovered this in 1968 from purely mathematical considerations, in a now-famous article titled, "Random Numbers Fall Mainly in the Planes."[21] However, it required a 3D plot, and one that could be animated or rotated interactively, to see the 3D structure of these RANDU numbers.

The left panel of Figure 9.14 shows a plot of successive triples from RANDU. Again, nothing unusual appears. Yet when this plot is rotated in 3D space, suddenly something totally *systematic* appears, as Marsaglia had predicted: all the points line up neatly on fifteen planes—a most decidedly nonrandom result. The result is somewhat like what you see when you drive past a field of corn: most of the time you see an apparently random arrangement of cornstalks; occasionally, a momentary view shows them neatly arranged in some systematic rows.

The simple algorithm for RANDU was ported from the IBM Scientific Subroutine library to other commercially available computer systems such as the Digital Equipment Corporation VAX. As a result of the wide use of RANDU in the early 1970s, many scientific results from simulation-based studies were seen as suspicious enough for many authors to redo their calculations with better random number generators to see whether their conclusions held.

The visual demonstration of the flaw in RANDU was slow in coming for the simple reason that computer technology for 3D displays and software for dynamic and interactive graphics did not exist until the early to mid-1980s. The next step was to take this from 3D into higher dimensions.

Travels in High-D Space

The principal idea for the display of data in three dimensions projected on a 2D image was well understood after Perozzo in the 1880s (see Chapter 8). The mathematics of rotating and otherwise transforming scenes in 3D space had also long been known. Even more, statisticians, starting with Karl Pearson and Harold Hotelling [1885–1973], fully developed the mathematical methods of approximating high-dimensional data in a smaller two- or three-dimensional space, analogous to shining a light on a 3D scene to see its shadow on a 2D surface. Statisticians could readily think of data in *n* dimensions, but they could not actually see or manipulate it, except in their mind's eye.

This all changed in 1973 with the development of PRIM-9, a project at the Stanford Linear Accelerator Center developed by Mary Ann Fisherkeller, Jerome Friedman, and John Tukey. PRIM-9 is an acronym for "Picturing, Rotating, Isolation and Masking in up to 9 dimensions." This was high-end stuff: the graphics hardware comprised a $400,000 graphic display system and a custom keypad controller. The computations to drive it were powered by an IBM 360/91 mainframe computer costing $500/hour.

Even now it is hard to describe in print how to explore and interact with a data space of nine dimensions using some expensive hardware and novel software.[22] It is something you have to demonstrate live. Before there were TED talks, Tukey did this in a movie now held in the ASA Video Archive.[23]

The keyboard controlled which variables were displayed at a given time: two main variables in a basic scatterplot, plus a third variable around whose axis the data were rotated in real time. Through projection and rotation, any arbitrary subspace of nine-dimensional data could be viewed dynamically.

Other controls provided isolation and masking, to select or view subsets of the data. Tukey's description of the added impact of these features was understated: "Together, choice of 2D view and freedom to rotate give more than 2D understanding. The dynamic effects of rotation allow us to see 3D structure unavailable in static views" (Tukey, PRIM-9 movie).

These ideas were revolutionary at the time, and they highlighted the central role of visual understanding in what Tukey dubbed *Exploratory Data Analysis* (EDA). At that time, research in statistics was most often couched in the language of mathematics: important results were stated in theorems, and numerical results were calculated to many decimal places. Tukey turned that around, to suggest that it was "far better an approximate answer to the right question, which is often vague, than an exact answer to the wrong question, which can always be made precise."[24]

With the advent of PRIM-9, new research methods for data analysis could be framed as problems in human-computer interaction. The results of an analysis of a complex problem could be explained and documented in a movie. More importantly, it established dynamic and interactive statistical graphics as a productive research area, one that attracted computer scientists and called for new techniques for interacting with data. Similar large-scale systems were developed at Harvard (PRIM-H) and the Swiss Federal Institute of Technology (PRIM-ETH) in the late 1970s. One example of scientific impact must suffice: a discovery of diabetes classification, aided by high-D visualization.

Diabetes Classification

Diabetes mellitus is a group of metabolic diseases in which people have high levels of blood sugar over a prolonged period. It was one of first human diseases identified: an Egyptian manuscript dating from circa 1500 BCE mentioned its primary symptom, "too great emptying of the urine." The role of the pancreas in diabetes was discovered in 1889 by Josef von Mering and Oskar Minkowski (brother of the famous mathematician Hermann Minkowski, who originated the spacetime diagram). But the mechanisms and forms of diabetes remained murky.

In a now famous story that led to a Nobel Prize[25] in 1923, Frederick Banting and Charles Best confirmed that it was insulin secreted by the pancreas that

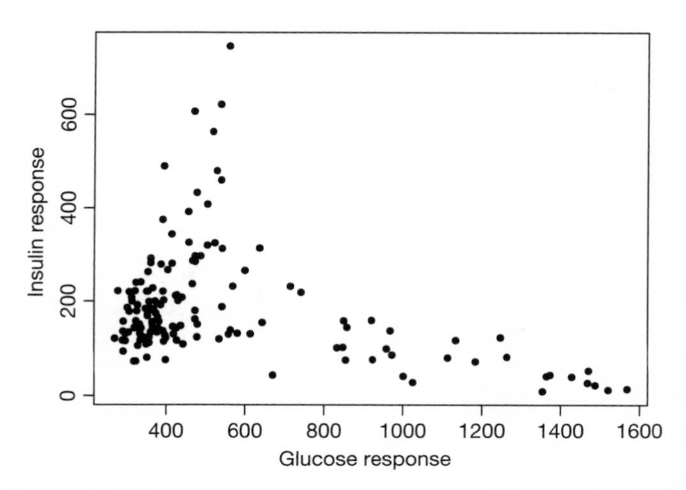

9.15 **Diabetes data:** Reproduction of a graph similar to that from Reaven and Miller (1968) on the relationship between glucose and insulin response to being given an oral dose of glucose. *Source:* © The Authors.

kept blood sugar in check. When they surgically removed the pancreas from a dog, it quickly exhibited the symptoms of diabetes and became progressively weaker. They rescued it from death by giving injections of an extract from the ground-up pancreas.

In 1968, the endocrinologist Gerald M. Reaven [1928–2018] and the statistician Rupert G. Miller [1936–1968] at Stanford University published a paper on the relationship between glucose levels in the blood and the production of insulin in normal subjects and in patients with varying degrees of hyperglycemia (elevated blood sugar level). They found a peculiar "horse shoe" shape in this relation (shown in Figure 9.15), about which they could only speculate: perhaps individuals with the best glucose tolerance also had the lowest levels of insulin as a response to an oral dose of glucose; perhaps those with low glucose response could secrete higher levels of insulin; perhaps those who were low on both glucose and insulin responses followed some other mechanism. In 2D plots, this was a mystery.

An answer to their questions came ten years later, when Reaven and Miller were able to visualize similar but new data in 3D using the PRIM-9 system. In a carefully controlled study, they also measured "steady state plasma glucose" (SSPG), a measure of the efficiency of use of insulin in the body, where large

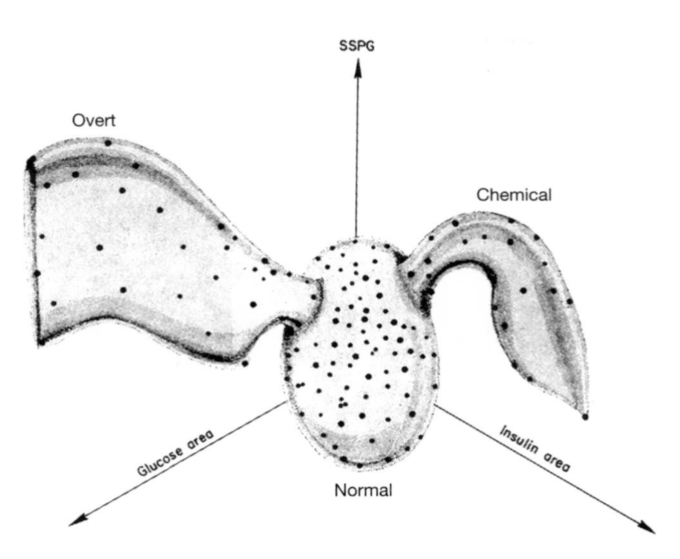

9.16 **PRIM9 view:** Artist's rendition of the data from Reaven and Miller (1979) as seen in three dimensions using the PRIM-9 system. Labels have been added by the author, identifying the three groups of patients. *Source:* G. M. Reaven and R. G. Miller, "An attempt to define the nature of chemical diabetes using a multidimensional analysis," *Diabetologia*, 16 (1979), pp. 17–24, fig. 1. Reprinted by permission from Springer Nature.

values mean insulin resistance, as well as other variables. PRIM-9 allowed them to explore various sets of three variables, and, more importantly, to rotate a given plot in three dimensions to search for interesting features. One plot that stood out concerned the relation between plasma glucose response, plasma insulin response, and SSPG response, shown in Figure 9.16. In a classic understatement of the importance of this finding, they said:

> The image . . . of 145 three-dimensional points was that of a boomerang with floppy wings and a fat middle. Given this visual perception of the three-dimensional relationship between the various metabolic variables chosen for examination, it seemed unlikely that the 145 subjects belonged to a single population. (Reaven and Miller, 1979, p. 18)

From this graphical insight, Reaven and Miller were able to classify the participants into three groups, based on clinical levels of glucose and insulin. The people in the wing on the left in Figure 9.16, who had high levels of fasting glucose, were considered to have overt diabetes. Those in the right wing were classified as chemical diabetics, and those in the central blob were classified as normal.

The visual understanding from this study was influential in defining the stages or classes in development of Type 2 diabetes. Overt diabetes is the most advanced stage, and is characterized by elevated fasting blood glucose concentration and classical symptoms. It had been thought that overt diabetes is preceded by the latent or chemical diabetic stage, with no symptoms of diabetes but demonstrable abnormality of oral or intravenous glucose tolerance.

However, it is clear from Figure 9.16 that the only "path" in the configuration of the data points from the chemical diabetics at the right to overt diabetics at the left goes through the region occupied by those classified as normal. This pattern suggested that the usual notion of a "natural history" of the disease—a smooth transition from chemical to overt diabetes—was wrong. Indeed, longitudinal studies showed that patients with chemical diabetes seldom developed overt diabetes. Instead, the evidence from this and other studies suggested that there were at least two general classes, "those who retain the ability to secrete substantial amounts of insulin and those who are insulin deficient."[26]

Gerald Reaven would continue to be among the foremost researchers on diabetes. In a 1988 Banting Lecture, he proposed that diabetes, high blood pressure, and male "apple-shaped" obesity have a common cause in insulin resistance and impaired tolerance to glucose. He called this "syndrome X" (it is now called "metabolic syndrome"), and this combination of symptoms has now been identified as a strong predictor of cardiovascular disease. Such was the power of interactive 3D graphics.

Hardware and Software

As a demonstration project, PRIM-9 suggested more than was ever accomplished with it. Creating dynamically rotated 3D graphics was painfully slow for larger data sets, except on the most powerful dedicated computers. The next steps in the development of ideas for visualizing data in time and space required specialized 3D graphics hardware and computer software to facilitate interactive exploration.

By the early 1970s, developments in computer animation and video games spurred electrical engineers to develop dedicated hardware chips to speed up graphics displays. These "large-scale integration" chips performed all the computations needed to represent objects in 2D space and make them move smoothly on a computer screen. The capability to produce moving images had been implanted in silicon circuits.

In 1972, Nolan Bushnell and Ted Dabney founded Atari and began the development of what would be called the Atari Video Computer System, a dedicated video game computer, with a joystick and other input devices. Arcade games, such as Pong, could be programmed and stored on ROM ("read-only memory") chips. The machine became the game, and it could be turned into a new game by plugging in a new chip.

In the decade following, computer engineers took this to a higher level, developing dedicated graphics processing units (GPUs), capable of high-speed graphic rendering in 3D, with hardware-backed "frame buffers" to support the computation of multiple scenes of an animated image, which produced a more realistic visual display.

In 1982, James Clark, an electrical engineer and computer scientist at Stanford University, founded Silicon Graphics, developing the idea of the "geometry pipeline," a combination of hardware and software to accelerate the animation and rendering of realistic three-dimensional images. By the mid-1980s, Silicon Graphics workstations, incorporating high-performance 3D graphics, became available to researchers in data visualization, which allowed the initial ideas behind PRIM-9 to be developed more widely. The original roots of this development in entertainment were not overlooked. By 1991, Silicon Graphics had become a world leader in the production of Hollywood movie visual effects and 3D imaging.

Software

Tukey and others had identified four key aspects of computer interaction in PRIM-9: *P*rojection and *R*otation provided the ability to see high-dimensional data on a 2D display, animated over time, with a third variable selected as the axis of rotation. *I*solation meant the ability to select interesting subsets of points, for example the clusters that Reaven and Miller would identify as overt and chemical diabetics. *M*asking was an early form of what is now called conditioning, showing successive subsets of the data either with different colors or plot symbols, animated over time, or in separate smaller views partitioned in space.

In the early 1970s, Alan Kay and other researchers at the Palo Alto Research Center (Xerox PARC) and elsewhere began to implement, in computer software, the ideas behind the graphical user interface (GUI)[27] for modern human-machine interaction (windows, icons, drop-down menus, drag-and-drop). Within a short time, these ideas became available in early personal computers (the Xerox Star, Apple Lisa, and Macintosh), which would open new possibilities for visual data analysis.

Over this and the next decade, research in interactive and dynamic data visualization accelerated in many places, with the development of custom software systems to support new methods for analysts to explore and visualize complex data. Carol Newton and later John McDonald, Andreas Buja, and others developed a new paradigm for these systems, based on what is now called "linked brushing in multiple views."[28]

Whereas PRIM-9 had only a single plot window, a key feature in later graphics software was the ability to show multiple plots in different panels of a graphic display. Brushing was the ability to select a region in any plot with a pointing device (now a mouse) and have the data in that region highlighted for further action (change color, hide, etc.). Linking was the main new idea here: whatever the viewer selected in one plot or data window would automatically be selected in all other windows.

Much of the early research and development used high-end graphics workstations, costing upward of $10,000 and the size of a small refrigerator. This began to change around 1978, with the introduction of desktop personal computers. Most notable was the Apple Macintosh, designed from the bottom-up with a sophisticated GUI as the basis for the entire operating system. Multiple windows, selection with a mouse, and drag-and-drop became built-in operations, available to all applications.

In 1984–1985, Andrew and David Donoho at Stanford University developed MacSpin,[29] a program for dynamic display of multivariate data, bringing the ideas of PRIM-9 and other novel interactive techniques to desktop computers. David Donoho demonstrated it at the American Statistical Association Meetings in 1986, showing that dynamic graphics had become as portable as a 25-lb. Macintosh. At around the same time, Paul Velleman, a student of Tukey now at Cornell University, developed DataDesk[30] (Figure 9.17), with similar aims but offering a more refined GUI for interacting with multiple linked views.

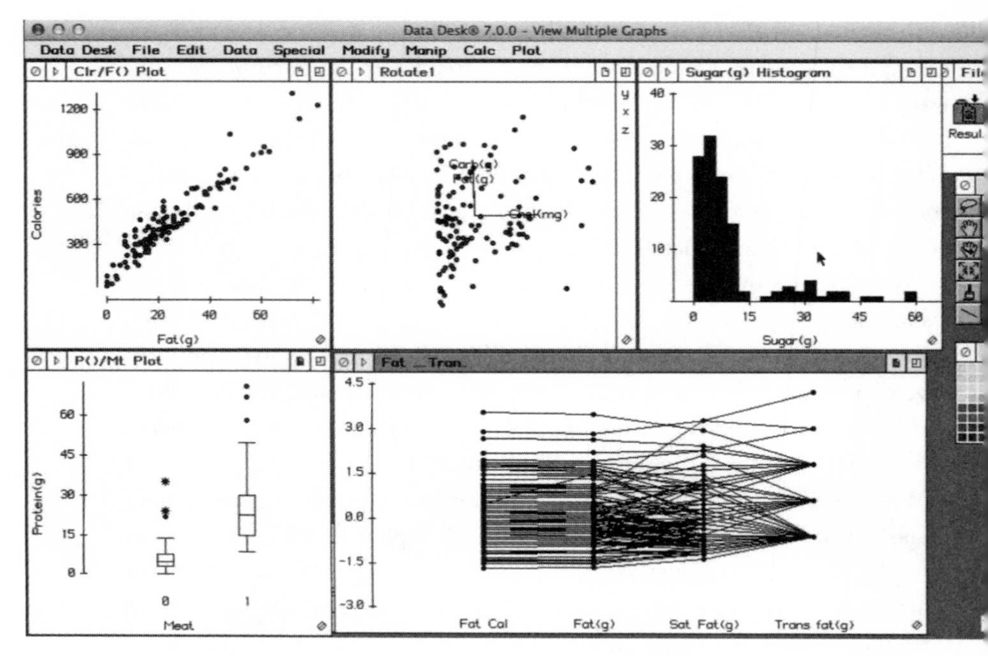

9.17 **DataDesk:** A multiwindow graphics system for the Apple Macintosh, providing dynamic graphics, linked plots and brushing. The plot windows show a scatterplot, a 3D plot that can be rotated around any axis, a histogram, a boxplot and a parallel-coordinates plot. A collection of plot control buttons is shown at the right. *Source:* DataDesk / YouTube.

Another powerful idea was introduced around this time. Imagine you are an explorer in a high-D world, searching for interesting features, like a creature from Flatland seeking to understand a cube, pyramid, or the solar system. In 1974, Jerome Friedman and John Tukey[31] tried to capture the visual insight from PRIM-9 in an algorithm, termed "projection pursuit," for finding revealing 2D projections of high-D data. By 1985 Daniel Asimov had developed what he called the "Grand Tour,"[32] which was designed to show a multivariate data set in an animation of 2D views by moving a plane on which the data points were projected according to some criterion of "interestingness." A cloud of points was interesting if it showed "clumps" (clusters), was "stringy," or showed prominent outliers. Such features could be quantified (with measures called "scagnostics"),[33] and such pictures could prompt researchers to think of possibilities they had not previously considered, as with the diabetes data.

Telling Stories with Animated Graphics

Development of software packages and libraries for animated graphics continued at an accelerated pace throughout the 1990s and continues today. Among these, one attracting wide general interest has been the moving bubble chart, which was designed by Hans Rosling [1948–2017], director of the Gapminder Foundation of the Karolinska Institute in Stockholm. The goal was to contribute to public debate about social issues of economy (income, poverty, inequality), health (life expectancy, infant mortality, HIV / AIDS), education (literacy, gender equality), the environment (pollution, water quality), and so forth by displaying public data using animated graphics in ways that could be readily understood by wide audiences.

The data initially consisted of time series for over 200 indicators derived from various sources (World Bank, OECD, World Health Organization, and similar organizations) for all available nations of the world; some of these had been recorded as far back as the early 1800s. Many are similar to variables that can be visualized as cartograms with WorldMapper—but only for a given point in time, and only for one variable at a time.

The moving bubble chart is essentially a scatterplot of two variables, x and y, depicting a third variable (z) as the size of a bubble symbol, which is colored according to a fourth categorical variable (k) and then presented as an animated movie showing changes over time (t). Using pull-down menus, any variables from the data set can be selected as x, y, z, and k. In addition, one or more countries can be selected from a list to highlight their changes over time.

In contrast to Playfair, who could only think in terms of two or more time series ($y_1, y_2, \ldots$) plotted against time, the moving bubble chart "liberated the x-axis from the burden of time."[34] More importantly, it provided Rosling with a vehicle to tell credible stories about a wide range of important social issues, which were illustrated with animated graphics showing five variables.

Figure 9.18 gives one example, which was used by Rosling in several video presentations, including one called "200 Years That Changed the World."[35] It shows three frames in the animation of the relation of life expectancy at birth to income per person at 100-year intervals. In 1815, the time of Napoleon's Russia Campaign, nearly all countries had a life expectancy under 40, and most had an income per person under $2000. By 1915, some countries had

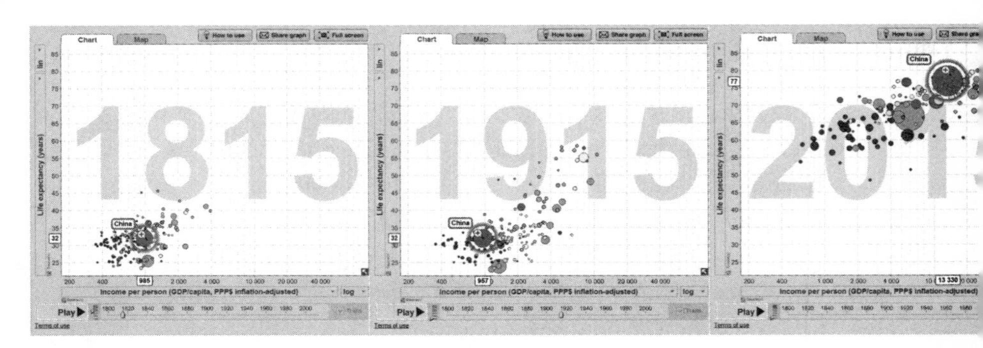

9.18 **Moving bubble chart:** Three frames of an animated sequence, plotting life expectancy against income per person for 142 nations from 1809 to 2015. Income is plotted on a log scale. The bubble area is proportional to total population and is colored by geographic region in the original. China has been selected to track its position over time. *Source:* Gapminder.

moved up to life expectancies in the 50s; most of these were ones whose income had increased, but most countries still had life expectancy under 40 and lower income. By 2015, most countries had moved up to a life expectancy in the 70s, which was associated with higher income, but there remained wide disparity in income, which was strongly associated with life expectancy.

Hans Rosling died in February 2017, but he left a large legacy of videos using global data on public health issues to inform public discussion on many of these topics in a way that just hits you between the eyes. Rosling was a consummate showman of visual explanation, sometimes a bit over the top, who used extravagant devices to make his points in a live presentation. Yet as visual stories of health and income inequality, they were powerful, persuasive, and memorable.

Implicit in Rosling's demonstration was that a critical component of powerful, persuasive, and memorable graphic presentations is the immediate connection a visualization can make between the data that make up a phenomenon and the consequential impact of those data on real human stories that they represent. We conclude our story with three truly evocative illuminations of this broader recognition of what data graphics paired with important data can accomplish, in our final chapter.

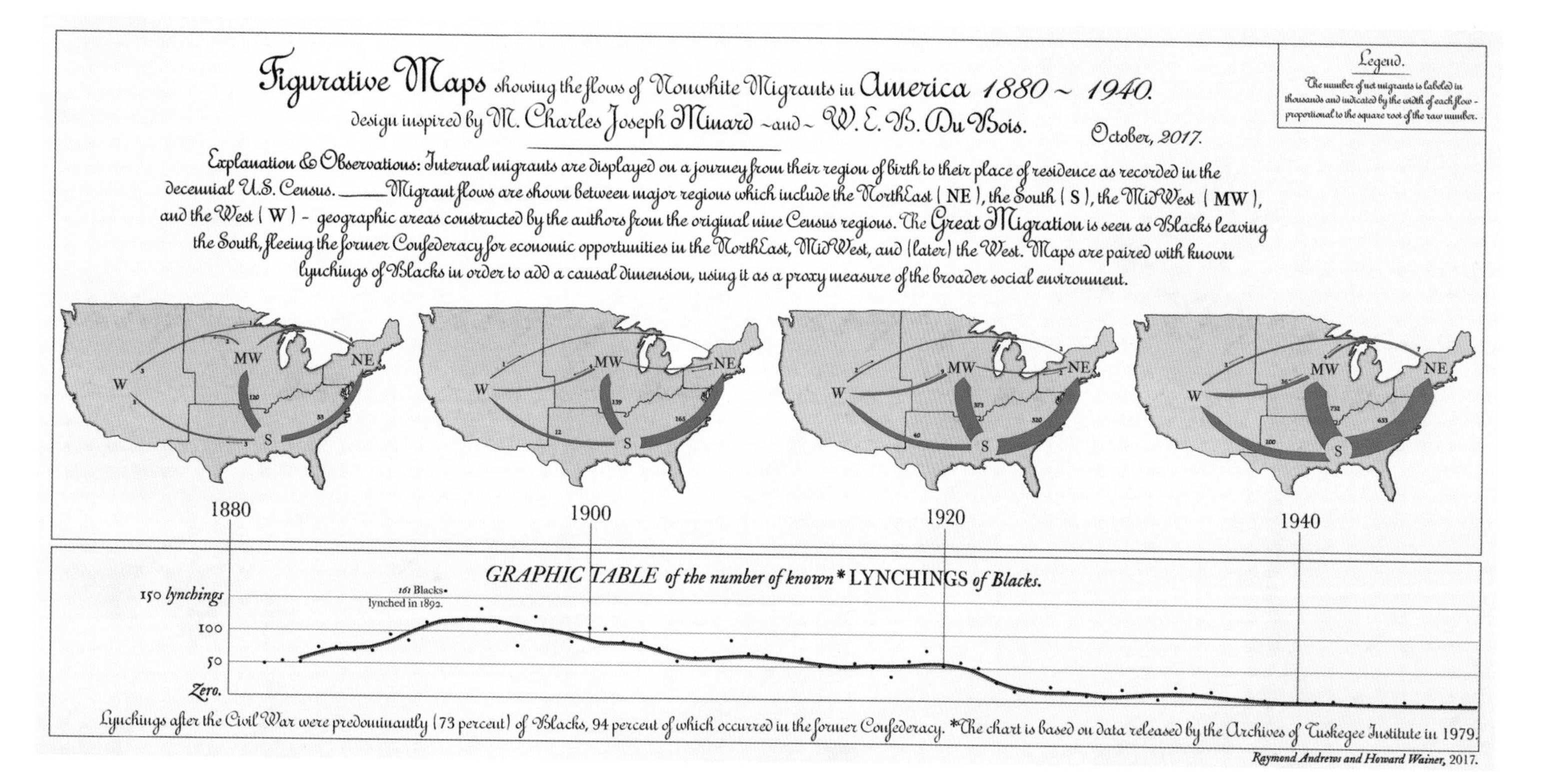

19 **Flow maps of migration:** Figurative maps showing the flows of non-White migrants in America, 1880–1940, using a design inspired by Minard and Du Bois. *Source:* R. J. Andrews and Howard Wainer, "The Great Migration: A Graphics Novel," *Significance,* 14:3 (2017), pp. 14–19, Figure 5. © The Royal Statistical Society.

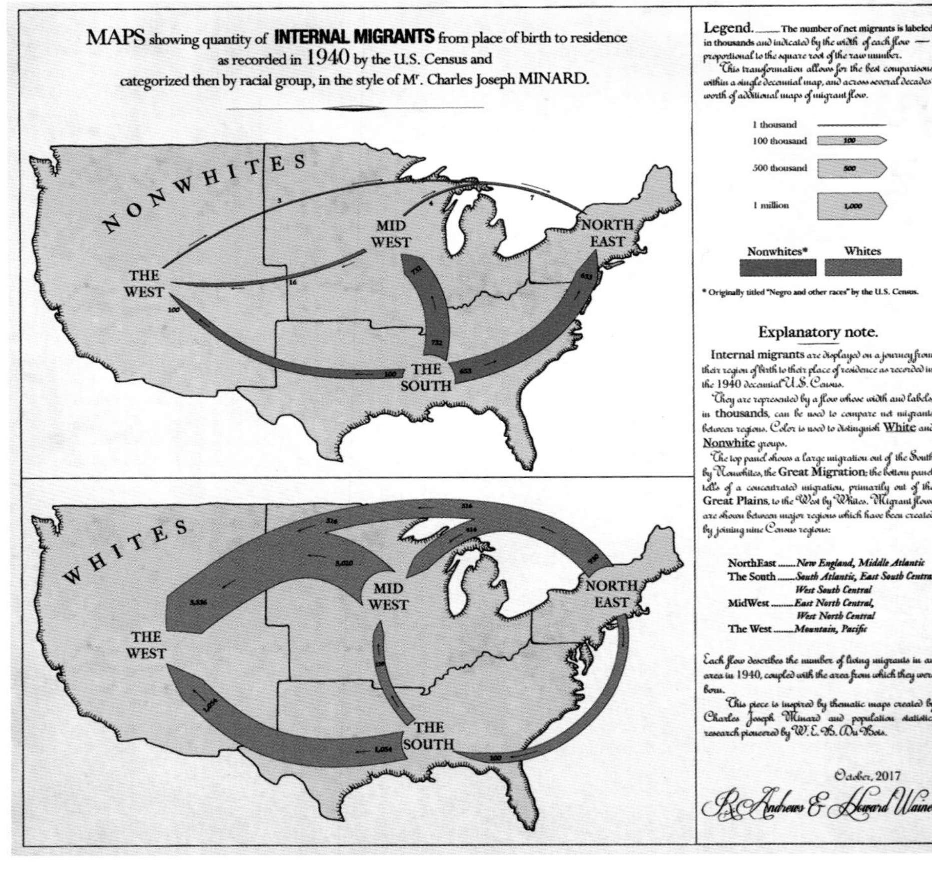

20 **Flow maps for 1940:** Maps showing the numbers of internal migrants by birthplace and place of residence, as recorded in the 1940 US Census, categorized by racial group, using a design inspired by Minard. *Source:* R. J. Andrews and Howard Wainer, "The Great Migration: A Graphics Novel," *Significance,* 14:3 (2017), pp. 14–19, Figure 6. © The Royal Statistical Society

10

Graphs as Poetry

Visual displays of empirical information are too often thought to be just compact summaries that, at their best, can clarify a muddled situation. This is partially true, as far as it goes, but it omits the magic. In the preceding chapters we have shown how the visualization of data is an alchemist that can make good scientists great and transform great scientists into giants.

In this coda we argue that sometimes, albeit too rarely, the combination of critical questions addressed by important data and illuminated by evocative displays can achieve a transcendent, and often wholly unexpected, result. At their best, visualizations can communicate emotions and feelings in addition to cold, hard facts.

The American poet Robert Lee Frost [1874–1963] famously pointed out that:

> Poetry is when an emotion has found its thought
> and the thought has found its words. (Frost, 1979, p. 283)

We can reverse the idea, inferring that one goal of the poet is to present the words that will engender the emotion.

The communication of an emotion is a difficult task that is a goal of many other means of communication. Surely music is the most obvious; in his *Ode to Joy* Beethoven transformed his emotions through his ideas and thence to music. The listener starts with the music and ends with exaltation. Visual artists have long followed this same path; Picasso transmitted his grief for the hundreds killed in the April 1937 bombing of the Basque town of Guernica during the Spanish Civil War in his mural *Guernica*, completed just two months after the bombing. For the receptive viewer, "the emotion is so lacerating that the next step beyond would be either insanity or suicide."[1]

Not all music is as transformational as *Ode to Joy*, nor are all paintings as evocative as *Guernica*; and most sequences of words are not poetic. But, happily, by Frost's definition, there have been poets in many media, and we can rejoice in their accomplishments. Goethe's observation that "Architecture is frozen music" (Goethe, 2018, p. 864) suggests we might find the soul of a poet in many architects. One undeniable example is Maya Lin, whose design for the simple chronological carving of 57,661 fallen soldiers' names into polished blocks of black granite yielded the heart-rending Vietnam Memorial in Washington, D.C.

We can include some data-based graphs in this elite company—graphs whose impacts on the viewer are so evocative to fully deserve the adjective "poetic." To reach this level requires a combination of data that have impact and a design of compact clarity. Reaching back to Frost's definition, this happens when an emotion has found its data and the data have found their design.

Let us begin slowly, by first introducing two graphic poems that might otherwise be overlooked because of their plain appearance. Then we describe a plausible result of a *gedanken* collaboration between two giants in their respective fields: Charles Joseph Minard, whose use of flow maps to communicate the horrors of war defied the limitations ordinarily associated with pen and ink, and W. E. B. Du Bois, who dedicated his long career to the difficult task of improving the lives and circumstances of African-Americans. He did this by broadly communicating more than a century of African-Americans' accomplishments despite their near-universal suffering under the yokes of slavery and racism.

Two Plain Graphical Poems

The conductor Ignat Solzhenitsyn, in a prelude to an all-Mozart concert, made the surprising observation that "Mozart's music is often underestimated because it is so beautiful."[2] Similarly, the poetry in some graphic designs is often unappreciated because those designs are so mundane. Few graphic artists have Minard's skill or his artist's eye, but we ought not be blind to poetry from others with less refined skills.

Young Men and Fire

After his retirement, Norman Maclean [1902–1990], a professor of Romantic poetry at the University of Chicago, wrote a book about the Mann Gulch fire, a forest fire that took place on August 5, 1949, in the Helena National Forest in

Montana. August is amid the worst part of the dry season, so when lightning struck the trees near the Meriwether Guard Station, it wasn't surprising that a fire broke out.

The fire quickly spread north to that portion of Mann Gulch that abuts the Missouri River. Once the fire was spotted, sixteen Forest Service smoke jumpers parachuted in to fight the blaze. The area around Mann Gulch is very rough country, and when the smoke jumpers were able to get close enough to judge the fire, they realized that it was beyond their ability to control and that their very survival depended on quick action.

The fire was moving rapidly up-gulch (northeast), so the smoke jumpers headed upslope, perpendicular to the fire's motion. They ran furiously, trying to get over the ridge that demarcated Mann Gulch ahead of the fire that was burning at their backs. But the land in that area is very steep, and the smoke jumpers needed to ascend to the ridge at an elevation of 4,700 feet that marked the boundary of safety. The charred bodies of thirteen of these young men were later found on this steep escarpment—within two hundred yards of the haven at the top of Mann Gulch.

Norman Maclean gathered the data to describe what turned out to be the final scramble for safety and constructed a graphical summary whose austerity contrasts with the tragedy it conveys (Figure 10.1). In Maclean's words,

> The convergence of fire and men in Mann Gulch offers itself as a tragic model for a graph, the modern scientist's favorite means of depicting what he wishes to present as clearly as possible. Drawn along axes of time and distance is one line depicting the course of the fire and one depicting the course of the men, and where there is convergence of the two, graphically speaking, is the tragic conclusion of the Mann Gulch story; the two lines converging to this conclusion constitute the plot. (Maclean, 1992, p. 269)

The rapid upturning of the dashed line conforms to the characteristic spread rate of fire, which increases in proportion to the square of the slope of the terrain—a stark contrast to the effect of steep slope on the swiftness of human ascent. It was very steep where the young men died.

As we study the graph, the character of the race between men and fire becomes more vivid still.

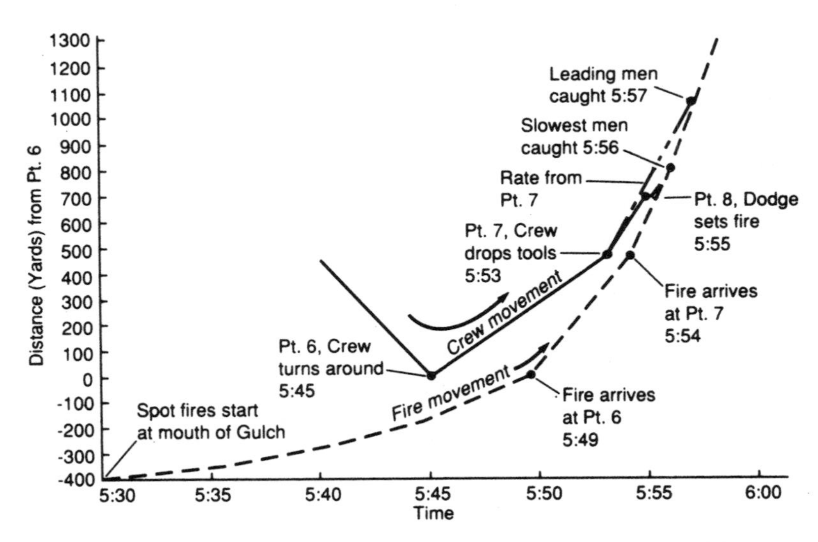

10.1 **Mann Gulch fire:** Distance and time graph of the estimated positions of the crew and the fire. Distances are estimated from the crew's turnaround at point (Pt. 6). The slopes of the lines indicate the rate of movement; the steeper the line, the faster the rate. The entire tragedy took less than thirty minutes. *Source:* Norman Maclean, *Young Men and Fire*. Chicago: University of Chicago Press, 1992, page 269.

> Along each line are numbers which are turning points in the race between men and fire, and if the lines are viewed as a race, the numbers mark off legs of the race, if they also have religious significance, they are stations of the cross, and if they have literary significance they mark off acts of drama, . . . but the acts are short, because modern wildfire allows no time for soliloquies. (Maclean, 1992, p. 294)

The Kovno Ghetto

During the Nazi occupation of Lithuania, the Nazis initiated a series of actions that resulted in the deaths of over 136,000 Jews. Initially, the Germans, in elaborately illustrated reports, meticulously documented these murders, but their production of records was reduced as the war continued. Indeed, in October 1943, SS chief Heinrich Himmler, in a speech to his subordinates, rationalized the minimized record keeping, saying, "This is an unwritten and never-to-be written page of glory in our history" (Braun and Wainer, 2004, p. 46). Nonetheless, the Nazis did create some documents that tabulated their triumphs, and others also kept records of their atrocities.

In many ghettos, Jewish leaders organized committees to keep official chronicles of daily life. Indeed, from the outset of the establishment of the Kovno Ghetto, Elkhanan Elkes, chairman of the Jewish council (the Ältestenrat), asked Kovno's Jews to write their own history as a legacy for future generations. Artists drew pictures, writers wrote stories, musicians composed music, poets wrote poems. They all used the skills and talent that they had to record both facts and emotions for posterity and for an audience that they feared they would not be able to meet directly. Ghetto residents without such artistic talent used whatever skills they had. Those with scientific training recorded data and presented them in a variety of formats, both tabular and graphical.

Figure 10.2 is one of those graphs, a traditional population pyramid, whose familiar simplicity and apparent banality belie its horrifying content. The bars represent the number of Jews in the ghetto by age, with the youngest at the bottom. The left side represents males, the right, females. The total length of each bar is the number of inhabitants of the ghetto of that age and sex in the beginning of October 1941. The shaded portion represents those still alive in November, less than two months later.

The Graphic Poetry of Charles Joseph Minard

Charles Joseph Minard [1781–1870] had a long and productive life. But official regulations forced him to retire from his long-held position at the French National School of Bridges and Roads on March 27, 1851, his seventieth birthday. Nevertheless, he continued working and teaching for the remainder of his life. The liberty he enjoyed in retirement allowed him to devote himself to projects he had begun earlier but that had been interrupted by the obligations of his job. Freed of other responsibilities, his development of new graphic forms and themes nearly doubled in rate for twenty years, and continued up to his death at age eighty-nine.[3]

In Chapter 7 we discussed Minard's justly famous map of Napoleon's Russian campaign (Figure 10.3), in which the initially mighty river of the 422,000 men of the French Grand Army crossing the Niemen River into Russia

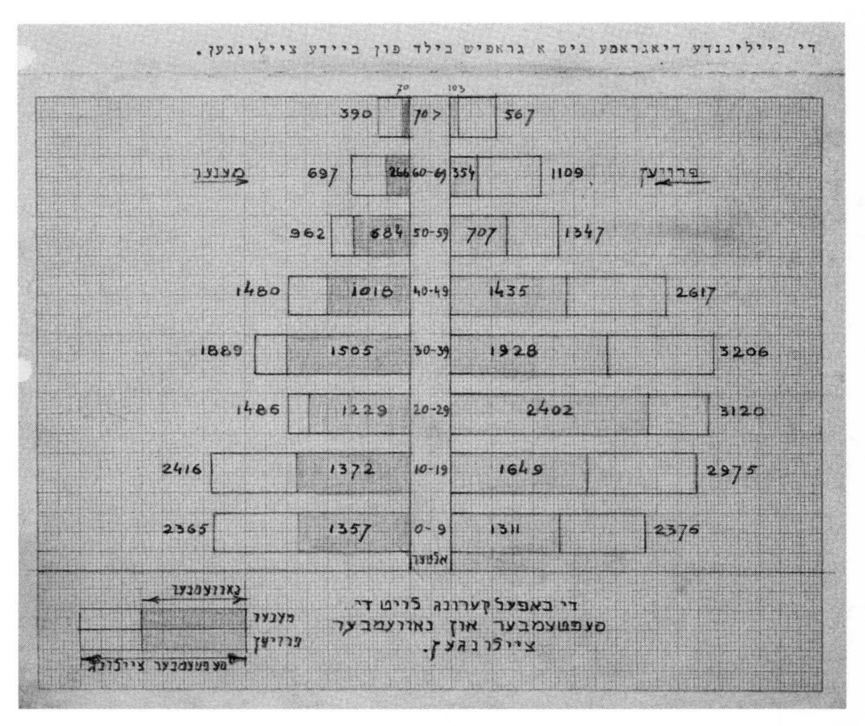

10.2 **Kovno Ghetto:** The population losses in the Kovno Ghetto due primarily to the "Great Action" of October 28, 1941. Males are represented on the left, females on the right. The shaded portion represents those still surviving in November. The central column indicates the age groups from 0–9-year-olds at the bottom to 70 and older at the top. *Source:* United States Holocaust Museum.

gradually diminishes, and is juxtaposed with the trickle of the surviving 10,000 returning at the end. When the French army reached Moscow in October of 1812 with only 100,000 men, they found it sacked and largely deserted. They turned back and marched across the steppes into the teeth of the Russian winter—Minard coupled the returning march with the descending temperature, shown in an accompanying panel below the map. He used the falling temperature as one plausible causal variable to connect the decline of the army's size with something in addition to the sniping of Russian soldiers. When this graphic was first published, the French physiologist and chronophotographer E. J. Marey[4] [1830–1904] was in awe, saying that it "defied the pen of the historian by its brutal eloquence" (Marey, 1885, p. 136).

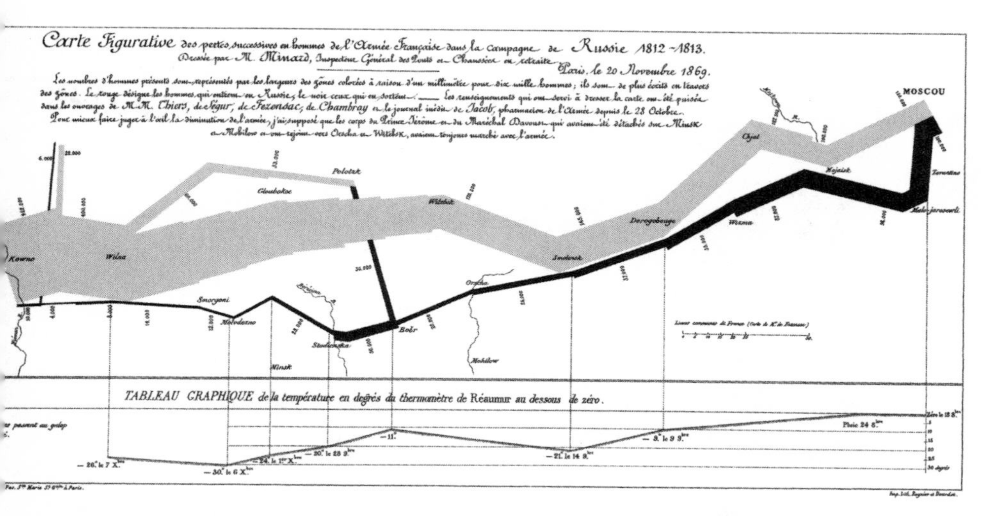

10.3 **Napoleon march graphic:** Charles Joseph Minard's narrative map of Napoleon's disastrous 1812 Russian campaign. The width of the gray "river" is proportional to the size of Napoleon's invading army; its black continuation shows the size of the returning army. *Source:* Reproduction courtesy of Archives, École Nationale des Ponts et Chaussées.

What was the source of inspiration for Minard's graphic poetry? This visual story of Napoleon's Russian campaign is noteworthy because it is the only known graphical portrayal of a national defeat by one of its citizens. It turns out that this was printed together with another graphic story of tragic military loss of life, two thousand years earlier (Figure 10.4).

The Carthaginian general Hannibal Barca fought the Second Punic War against the Romans from 218 to 201 BCE. His crossing of the Alps in 218 BCE is considered one of the most celebrated achievements of any military force in ancient warfare because it allowed him to bypass Roman land garrisons and naval dominance and take the war directly into the heart of the Roman Republic.

Figure 10.4 is Minard's visual story.[5] Hannibal's campaign begins in Iberia, in southern Spain. His army, consisting of over 100,000 soldiers and numerous war elephants is shown to cross Spain and then southern France. But, from October to December 218 BCE, the snow in the Alps and the steep descent on the Italian side proved treacherous; all the elephants, most of the pack animals, and many men perished. Hannibal's army emerged in the Po Valley in Italy with a mere 26,000 troops.

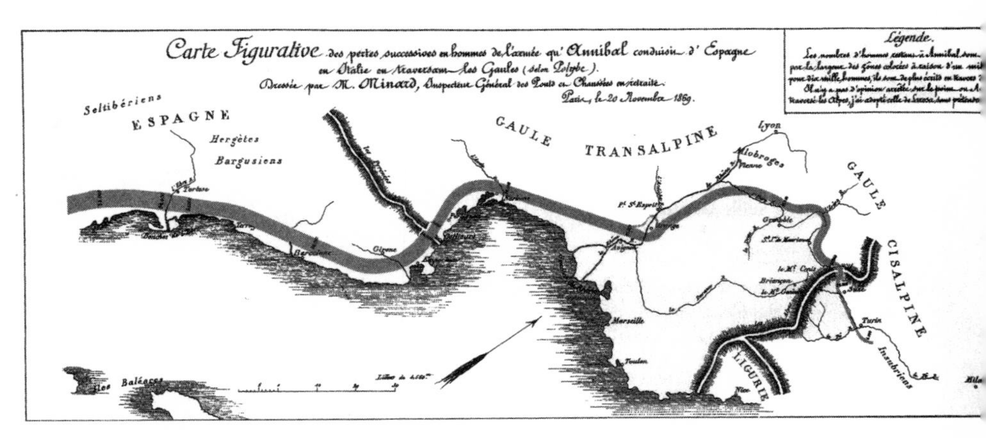

10.4 **Hannibal graphic:** Minard's depiction of Hannibal's army, as it treks across Spain and France on a brilliant military campaign, but then suffers huge losses attempting to cross the Alps. *Source:* Reproduction courtesy of Archives, École Nationale des Ponts et Chaussées.

The loss of life was not nearly as dramatic as that suffered by Napoleon's troops in Russia, but the map does draw visual attention to the relatively large loss as Hannibal crossed the Alps. Together, the maps of these two campaigns provide a visual lesson for historians and generals. These might have been titled "Some things to avoid in planning a military campaign," but Minard's titles are more direct: "Figurative map of the successive losses of men in the army" (Carte Figurative des pertes successives en hommes de l'Armée Francaise dans la campagne de Russie 1812–1813).

Minard's personal history makes the source of his inspiration clear. As a young engineer in Anvers in 1813, he had witnessed the horrors of war in a siege by the Prussian army. In his eighty-ninth year, 1869, he was deeply troubled by what he foresaw as an inevitable new war with Prussia and the havoc that would wreak on the Second French Empire. The two "figurative maps" of Napoleon and Hannibal were published on November 20, 1869.

Minard's fears proved correct. The Franco-Prussian War began on July 19, 1870. France was ultimately defeated and, following a long siege, Paris fell on January 28, 1871. Minard foresaw this too and, though now infirm and requiring crutches, he left Paris for Bordeaux on September 11, 1870, leaving behind nearly all his books and papers. He carried with him only a few works-in-progress, but these evidently have been lost. Sadly, six weeks later he contracted a fever, and on October 24, 1870, he died.

Very little of Minard's personal life is known;[6] the main historical source is the obituary written by his son-in-law, Victor Chevallier in 1871.[7] He says, "Finally, . . . as if he could sense the terrible disaster that was about to disrupt the country, he illustrated the loss of lives that had been caused by . . . Hannibal and Napoleon. . . . The graphical representation is gripping; . . . it inspires bitter reflections on the human cost of the thirst for military glory" (Chevallier, 1871, p. 18). It may well be, for this reason, that Minard's most famous graphic defied the pen of the historian, and these two graphic stories have become the iconic examples of Minard's graphic poetry.

Using Graphs in a Narrative Argument

Using a series of graphical displays to construct an empirical narrative was not common in Minard's time. More often, a single display was used to convey a single idea. Napoleon's march (Figure 10.3), as breathtakingly marvelous as it is, is still a short story, with the limitations commonly associated with that form. Embedding a sequence of displays in an empirical narrative has been too rarely used. But when used effectively, this approach forms a memorable image that has remarkable impact.

Minard himself sometimes used comparative graphs to tell a story. As one example, he prepared a sequence of plots to show how the Union blockade of Confederate ports during the American Civil War affected the supply of cotton to English mills. In Plate 17 (shown previously as Figure 7.5), we reproduce his three-part plot (for 1858, 1864, and 1865), which made it clear that before the war the southern US states were the principal supplier for British mills. But when that supply was effectively cut off by the blockade, the mill owners replaced it with cotton from India and Egypt, which continued, at a slightly diminished level, even after the war had ended.

Arguably, the most famous (and perhaps the first) use of a sequence of displays intermingled with a verbal explanation was Leonardo's magnificent tale (Figure 10.5) of how a human fetus is carried in a womb. It challenged Galen's 1,400-year-old description of the human womb as bicameral (having two chambers) to carry multiple births.

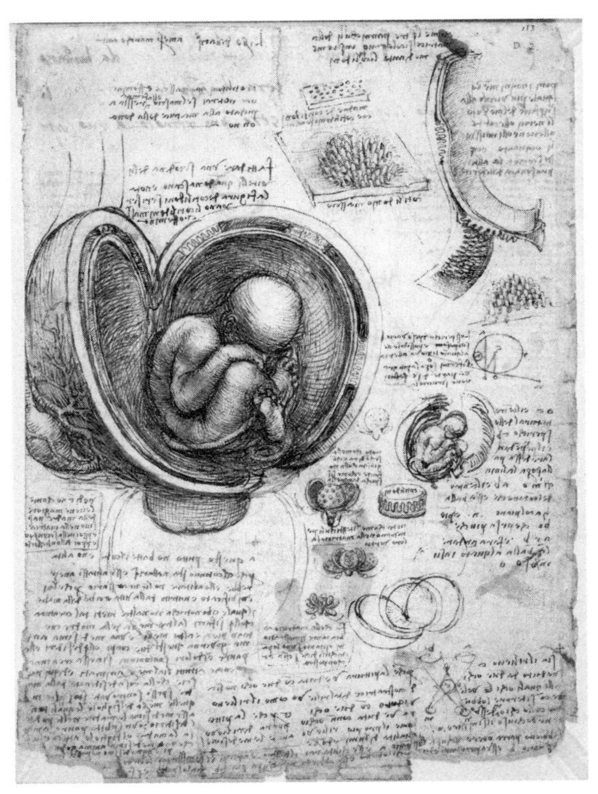

10.5 **Da Vinci's notebook:** Leonardo's graphic story of a fetus told with words and incomparable pictures that disproved Galen of Pergamum's claim about women's bicameral uterus, which had persisted unchallenged for 1,400 years. *Source:* Wikimedia Commons.

W. E. B. Du Bois

William Edward Burghardt (W. E. B.) Du Bois was born in Great Barrington, Massachusetts, on February 23, 1868, shortly after the end of slavery in the United States. He died in Accra, Ghana, on August 27, 1963. During the course of his ninety-five years he accumulated an extraordinary list of accomplishments. He earned a PhD from Harvard—the first African-American to do so. He was by turns a sociologist, historian, civil rights activist, and author. His many books included *The Souls of Black Folk*, *The Suppression of the African Slave Trade*, *Dusk of Dawn*, *Darkwater*, and

The Talented Tenth. He also had a keen appreciation of the power of data in a narrative.

Harold Geneen, the enormously successful CEO of ITT Industries, Inc., was a man dedicated to using empirical information. He pointed out that "when you have mastered numbers, you will in fact no longer be reading numbers, any more than you read words when reading books. You will be reading meaning."[8]

It is natural to ask how Geneen in 1985, and DuBois a century earlier, were able to look at numbers and extract meaning? Certainly the consensus then, and now, is that the extraction of meaning from the huge tables of numbers that might carry that meaning is wearisome to the eye and difficult for the brain; as the American economists Arthur Briggs Farquahr and Henry Farquahr so memorably put it 1898, it is a task akin to extracting sunbeams from cucumbers. We can plausibly conclude that DuBois was best able to read meaning from numbers after they had been transformed to visual displays. In his remarkable success at this, he contributed to the development of the modern approach to constructing a narrative argument based on evidence through his use of a series of graphic displays. In this way he is a direct, lineal descendant of Leonardo and Minard.

In 1900 Du Bois collaborated with Booker T. Washington on "The Exhibit of American Negroes" at the Paris World Fair (the *Exposition Universelle Internationale*). It included 400 patents by African-Americans as well as 200 books with African-American authors. As well, a large number of facts about African-Americans were woven into a memorable narrative by transforming them into data graphics.[9]

The nearly sixty graphs and thematic maps spanned a wide range of characteristics of African-Americans and their lives. This section of the exhibit was titled "A series of statistical charts illustrating the condition of the descendants of former African slaves now in residence in the United States of America."

Du Bois began his story with a graph dramatically showing the profound effect of Lincoln's Emancipation Proclamation (Figure 10.6). It plots the balance of enslaved versus free African-Americans from 1790 until 1870. From this we learn that the proportion of free African-Americans during slavery hovered around 12 percent of the total African-American population until January 1863, when Lincoln's Emancipation Proclamation[10] was the key to the dramatic change shown.

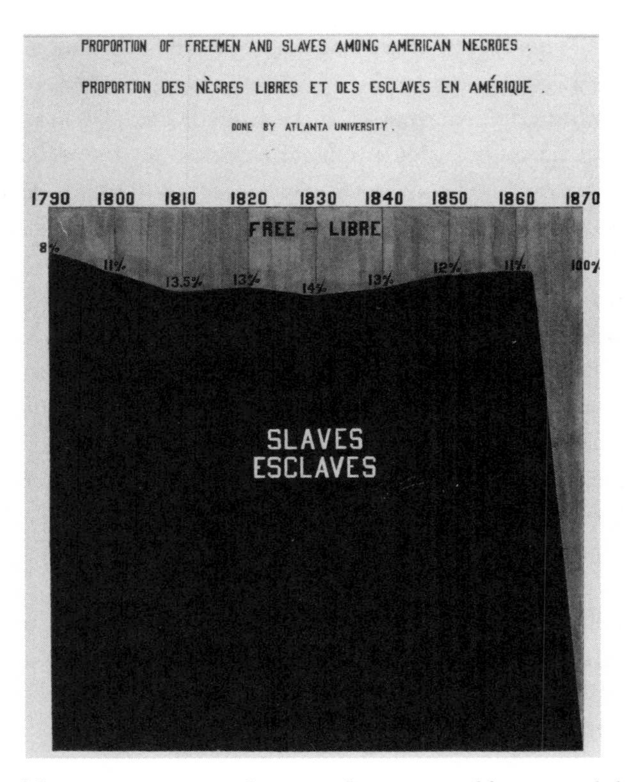

10.6 **Effect of the Emancipation Proclamation:** "Proportion of freemen and slaves among American Negroes, from 1790 to 1870." *Source:* Library of Congress, Prints and Photographs Division, LC-DIG-ppmsca-33913.

Having established the status of African-Americans in the United States, he then used a simple bar chart (Figure 10.7) to show that there were 7.5 million African-Americans in 1890 and also the exponential growth of that population over the prior 150 years. Although a simple time-series line graph in the style of Playfair might have been used, Du Bois chose the format of a horizontal bar chart in the style of Minard to also show the exact numbers.

To put the size of the US population of African-Americans at that time into a global context, Du Bois presented a plot showing the outlines of ten other countries in which their size was proportional to their total population; in the center was a map of the United States, whose size was proportional to the African-American population (Figure 10.8). Study of this plot quickly reveals

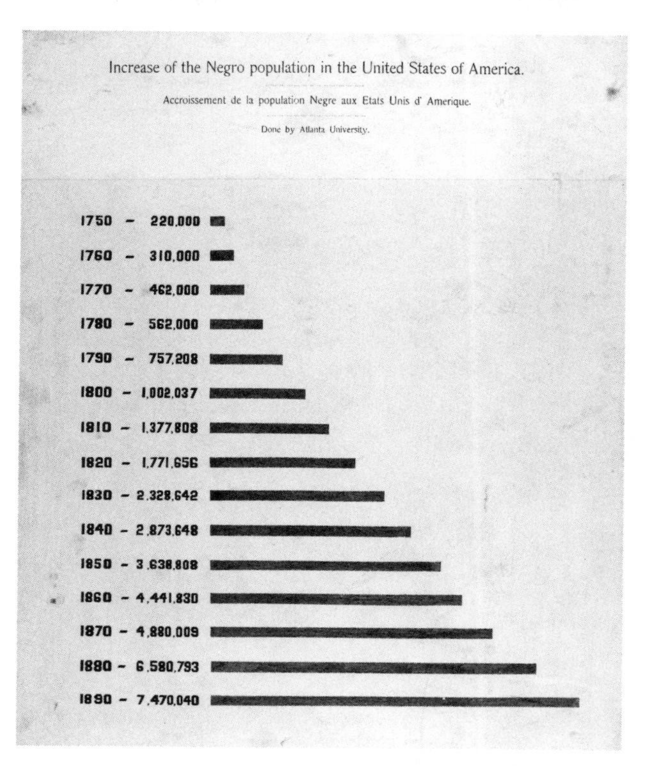

10.7 **Du Bois's bar chart:** Increase of the Negro population of the United States from 1750 until 1890. *Source:* Library of Congress, Prints and Photographs Division, LC-DIG-ppmsca-33901.

that the number of African-Americans is greater than the entire populations of Australia, Norway, Sweden, Holland, Belgium, Switzerland, and Bavaria.

Du Bois used this same graphical form to compare the size of the African-American population with the white population of the United States (Figure 10.9), augmenting the plot with proportional figures, showing that the African-American portion of the US population had been diminishing over time, while the total population grew.

He made the differential rates of growth explicit in a well-designed line plot (Figure 10.10), showing that, even though the African-American population had been growing exponentially, so too had the rest of the United States, and the rate of the latter's growth was larger. In the style of Playfair, he used text labels to show important historical events, such as the suppression of the

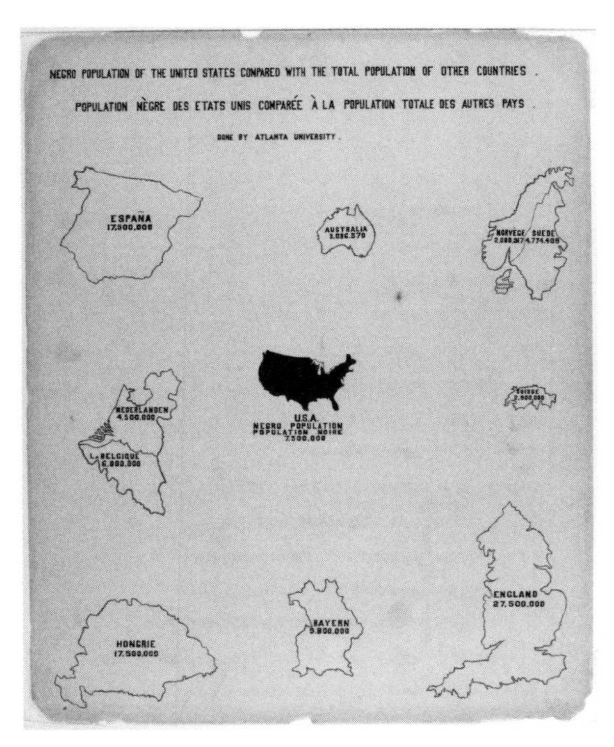

10.8 **Cartogram maps:** "Negro population of the United States compared with the total population of other countries." *Source:* Library of Congress, Prints and Photographs Division, LC-DIG-ppmsca-33903.

slave trade in 1810, to the Emancipation Proclamation in 1865. In the style of Minard, he also showed the numerical values of percentage increase per decade on each curve.

Du Bois used these displays to answer fundamental questions of existence: How many of us are there? Are we increasing or decreasing in number, how fast (both absolutely and in comparison to Whites)? The next obvious question is Where do we live? He addressed this in Plate 18, which showed dramatically the heavy concentration of African-Americans in what had been the states of the Confederacy, but also signs of a movement out of the Deep South into the Northeast.[11]

Du Bois continued with fifty-two other plots, expanding and enriching the narrative. This fact-based description was made possible by the 1870

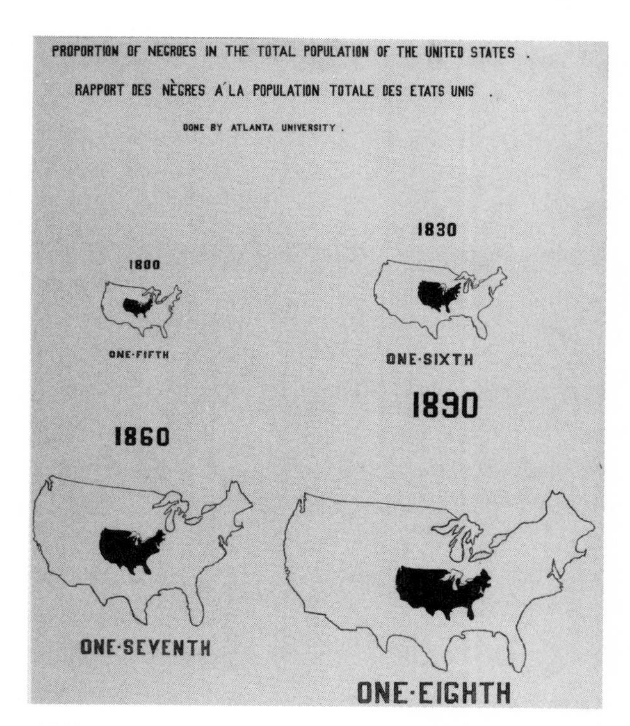

10.9 **Cartogram maps:** "Proportion of Negroes in the total population of the United States." *Source:* Library of Congress, Prints and Photographs Division, LC-DIG-ppmsca-33904.

expansion of the US Census, which, for the first time, included the African-American citizens in the national accounting.

Absent, however was any sort of dynamic representation of the gigantic migration of African-Americans out of the South that took place after Reconstruction and continued for the first half of the twentieth century. This phenomenon could have been beautifully communicated with a collaborative effort between Du Bois and Minard.

The Great Migration

As seen in Figures 10.7 and 10.10, there were 7.5 million African-Americans in 1890, the great majority of whom lived in the rural South of the United States. Between 1916 and 1970 more than 6 million migrated to the industrial

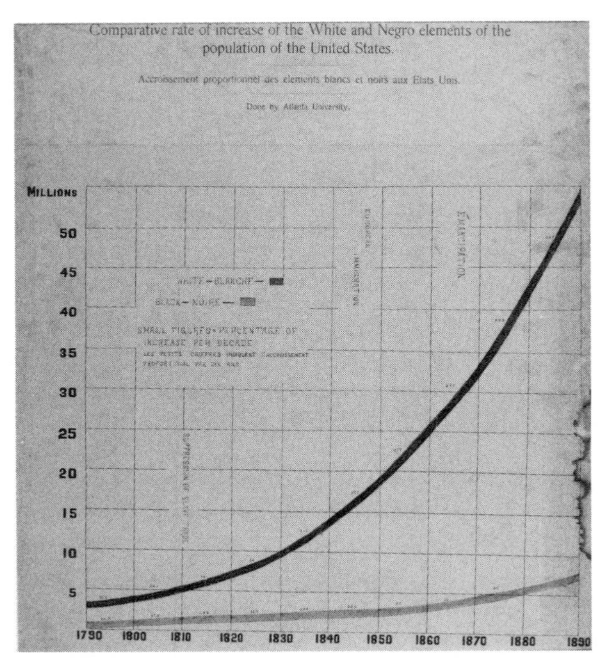

10.10 **Relative growth line chart:** "Comparative rate of increase of the White and Negro elements of the population of the United States." *Source:* Library of Congress, Prints and Photographs Division, LC-DIG-ppmsca-33902.

North and West. Historians such as Nicholas Lemann have dubbed this the Great Migration, commemorating it as the largest peacetime movement of people within a country in history.[12]

Such a movement cries out for description and explanation. The biggest initial question is, of course, "Why?" But before one can tackle such an interpretative question, one faces the descriptive questions: "From where to where?" "When?" and How many?" These questions are fully answered with data, indeed, the sorts of data that are routinely and rigorously gathered by the US Decennial Census. But extracting geographic structure from the entangled tables of numbers describing it brings to mind the cucumbers of the brothers Farquahr. And communicating the messages contained in these census tables beyond the eyes of experts is far better done visually. But how are we to transform the tables to graphics?

To answer this question we adjoin the life work of the two giants whose work forms the center of this chapter: Du Bois's passion for improving the lives of African-Americans and widely communicating their accomplishments, and Minard's genius for transforming complex data into compelling pictures that impacted both the intellect and the emotions.

Although their long lives overlapped by but two years (and even during that brief period they were separated by language, culture, and an ocean) the work and goals of Minard and Du Bois were complementary. In the balance of this chapter we demonstrate of how a blending of their work yields an evocative narrative of such awakening power that it can rightfully be described as poetic.

The census has always been tasked with counting all residents of the United States; in 1863 its scope was expanded to include those African-Americans who had been emancipated from slavery. Thus, the 1870 census, just after the Civil War, included counts of African-Americans for the first time.[13] These census data provided counts, by race, place of birth and place of residence, making it straightforward to study migration patterns of people.

The newly freed African-Americans faced impossibly difficult times. They were desperately poor, largely illiterate, and living in a region whose plantation economy offered little chance for improvement. Most worked as sharecroppers, tenant farmers, or farm laborers. In addition to economic woes, the post-Reconstruction South was racist and violent: Jim Crow laws instituted de jure racial segregation, and racist organizations like the Ku Klux Klan, the White League, and the Red Shirts made terror a part of daily life for African-Americans living in the post–Civil War Confederate South. Thus, it isn't surprising that large numbers of African-Americans opted to migrate to the North for greater opportunities. World War I created a huge demand for workers in northern factories; northern railroads needed workers so badly that they provided free train passes for thousands of southern Blacks.

A Gedanken *Collaboration*

Let us now imagine a collaboration between Du Bois, who chose the topic on which we focus our attention, the census (which supplied the data), and Minard, who invented and perfected the graphical method that would allow us to see the character of the Great Migration, which, over the fifty-five year span from 1915 until 1970 saw more than 6 million African-Americans leave the South.[14] We suspect that Minard would have appreciated this graphic challenge because it parallels his 1862 powerful graphic story, which showed the pattern of worldwide emigration in a single flow map.[15]

Plate 19 shows a compound plot of four flow maps drawn in the style of C. J. Minard. The four panels are spaced twenty years apart, showing the movement of non-Whites (the census designation, which consisted primarily of African-Americans). The arrows show the net migration (outflow *minus* inflow), and the widths of the arrows are proportional to the number of migrants. The panel at the bottom depicts the annual number of lynchings of African-Americans; 94 percent of these were in the former Confederacy. We are using this variable in the same way that Minard used temperature in his map of Napoleon's march—as one possible causal variable.

The leftmost panel of Plate 19 (for 1880) shows the beginning of the post-Reconstruction exodus. Most of the migrants left the South for the industrial cities of the North and Northeast (Chicago, Detroit, Pittsburgh, and Washington, D.C.). The number of lynchings increased, peaking in 1890, and in 1900 the migration more than doubled. The need for labor expanded during World War I and is manifested by the further increases shown in the 1920 panel, when 454,000 people left the South in the decade 1910–1920, and the 1940 panel, when another 1,500,000 left during the Depression and the buildup to World War II. Another principal cause was the great Mississippi Flood of 1927, in which more than 200,000 African-Americans were displaced from their homes on the lower Mississippi River.

The Great Depression affected huge numbers of people, both White and Black, many of whom lost their homes (particularly in the Dust Bowl states) and left for a better life elsewhere. This is shown for African-Americans in the top panel of Plate 20 (an enlargement of the 1940 panel of Plate 19). The bottom panel (1940 data for Whites) provides a dramatic contrast. Although African-Americans fled the South in the decades of the Great Depression in much the same patterns as they did in the decades before and after, the Whites were often farmers whose economic lives were ruined by the Dust Bowl catastrophe (think of the Joads in Steinbeck's *Grapes of Wrath*, who piled all their belongings onto the family truck and headed West to California).

Conclusion

So we beat on, boats against the current, borne back ceaselessly into the past.

—Nick Carraway's haunting last line from
F. SCOTT FITZGERALD, *The Great Gatsby*

This chapter, in a very real sense, provides a coda for the entire book. We have found that whenever we face a new problem it is always wise to look backward first; this provides the benefit of helping us to understand the thinking of those who have preceded us. The hubris that drives some to believe that we are wiser than our forebears rarely yields a happy result. This entire book is driven by the belief that we are best prepared to move into the unknown future by better understanding the past. In addition to illuminating the past, we wished to draw attention to the extraordinary power of visual depiction of quantitative phenomena to communicate efficiently both facts and emotion—how a graph can be a poem.

To illustrate these points, we have provided a portrait of the Great Migration of African-Americans from the South of the United States in the century following their emancipation from slavery. Sensible space limitations forced us to make some simplifications. In doing so we found that Mies van der Rohe's dictum "less is more" was spectacularly true and faced some of the same choices for level of detail versus fidelity that Du Bois and Minard encountered in their work.

Looking at decade-by-decade plots added little that was more clearly shown by collapsing to twenty-year periods; showing migration destinations to the nine census regions, rather than the four super-regions we constructed, added more noise than structure; including a 1960 plot that contains the final stage of the exodus, when almost 3.5 million people left the South for the North and the West adds quantity to what we have already shown but does not change the overall message. Thus, we believe that this summary is both accurate and complete, and it provides a suitable continuation of the task that W. E. B. Du Bois began more than a century ago.

Du Bois's graphical narrative showed facts about African-Americans; how they started out and what they became. We added some details of how they got there and where they came from. The tool we chose to do this we snatched from the past, mining the ideas of the great C. J. Minard. We shamelessly borrowed his metaphor of the flow of goods or people across a geographic background. Minard's adjoining of the delicate curve of declining temperature with the rapid shrinkage of Napoleon's army in their death march across the Russian steppes in winter 1812 forces the viewer to share the great sadness that enveloped all of France. Our use of the ancillary variable of lynchings was meant to generate a similar emotion, as one empathizes with the terror that African-American southerners must have felt as they made the decision to leave home and seek to rebuild their lives elsewhere.

It doesn't take an excessive amount of imagination to connect this same methodology to the Jewish diaspora from Nazi Germany, the Hebrews from Pharaoh's Egypt, peasants during the eastern European pogroms, the Cherokees' Trail of Tears, or the Bataan Death March. By memorializing the eighteenth century British expulsion of the Arcadians from eastern Canada in his epic poem *Evangeline*, the American poet Henry Wadsworth Longfellow, captured the popular imagination and garnered enough sympathy for the Arcadians to induce the British government, on July 11, 1764, to permit displaced Arcadians to return home. Such is the power of poetry.

We will leave it to others to prepare graphic accounts so that we can better understand and remember the facts and the emotions of these tragic events. For poetic data visualizations, combining both the communication of facts that can be learned with emotions that can live in your heart forever, can help provide a bulwark against future horrors, for "*there is a magic in graphs. The profile of a curve reveals in a flash a whole situation—the life history of an epidemic, a panic, or an era of prosperity. The curve informs the mind, awakens the imagination, convinces.*"[16]

Learning More

This section contains some links and references to further reading and topics that you might wish to explore. They didn't fit directly in the narrative of a given chapter but are useful nonetheless.

Chapter 2

- This chapter draws on material from Friendly et al. (2010). The supplementary web page for this paper, https://datavis.ca/gallery/langren/, gives the historical sources, including earlier versions of Figure 2.1 contained in various letters, translations of *La Verdadera*, and the text of van Langren's cipher.
- Van Langren's cipher is still one of the most elusive unsolved problems in cryptography. If you like the history of this topic, you might enjoy the 2017 book by Craig Bauer, *Unsolved!*
- A popular history of the problem of longitude by Dava Sobel (1996) focuses on the work of the clockmaker John Harrison who, only with considerable difficulty, was finally perceived to have solved the problem to sufficient accuracy to be awarded a prize by the Longitude Commission.
- The story of the mapping and naming of lunar features did not begin or end with van Langren. Ewen Whitaker's *Mapping and Naming the Moon* (2003) gives a comprehensive history and devotes a chapter to van Langren, including other versions of the 1645 lunar map.
- In the period after van Langren, other early things that could be called graphs were often produced by devices that recorded some phenomenon like temperature or barometric pressure with a pen on

moving paper or a drum. Robert Plot's chart of barometric pressure (Figure 1.4) is one example. Hoff and Geddes (1962) give a detailed history of this early history with many fine illustrations.

Chapter 3

- The wider story of the roles of empirical observation and data in the intellectual development of science in the eighteenth and nineteenth centuries is well told by Ian Hacking in *The Taming of Chance* (1990).
- The story of Guerry's *Moral Statistics of France* is told in greater detail in Friendly (2007). This article also describes his later work and relates his data and questions to modern methods of statistics and graphics. A web supplement, https://datavis.ca/gallery/guerry, provides resource material, and Guerry's data have been made available in the R package Guerry, https://CRAN.R-project.org/package=Guerry.
- Very little of Guerry's personal life or family history was known until recently. A brief biography is available (in French) in Friendly (2008c) and an English version is linked on the datavis.ca web site mentioned previously.
- Along the way, Guerry tabulated so much data that he invented a mechanical calculator, an *ordonnateur statistique*, to help in this work. The history of this device, which perhaps was the first special-purpose statistical calculator, is described in Friendly and de Saint Agathe (2012).
- In England, Joseph Fletcher comes closest to Guerry in his pursuit of the relations among moral variables and his use of thematic maps to display such data. Cook and Wainer (2012) describe these contributions.
- The most comprehensive treatment of the early development of thematic mapping is still Robinson (1982). Palsky (1996) gives a detailed history of quantitative cartography in the nineteenth century (in French), and Friendly and Palsky (2007) provide a history of thematic maps and diagrams designed to explore the connections between graphic images and scientific questioning. Delaney (2012) provides a richly illustrated overview of some landmark developments in this area.

Chapter 4

- Steven Johnson's *The Ghost Map* (Johnson, 2006) is a compelling popular description of the background for the cholera outbreaks in London and the roles that Snow and others played in uncovering evidence and tracing the initial outbreak to the "index case," Frances Lewis, a five-month-old child residing at 40 Broad Street, adjacent to the pump.
- *Disease Maps* by Tom Koch (2011) traces a history of the medical uses of cartography to understand the outbreak and transmission of disease.
- Scott Klein, a data journalist at ProPublica, presents an interesting look at how journalists at the New York *Tribune* covered an outbreak of cholera in New York City in September 1849, using a time-series line graph of cholera deaths on its front page. See https://www.propublica.org/nerds/item/infographics-in-the-time-of-cholera.
- A project initiated by Lynn McDonald, *The Collected Works of Florence Nightingale*, comprises all her available surviving writing (letters, articles, pamphlets, etc.) in sixteen volumes. An online catalog is available at https://cwfn.uoguelph.ca/.
- WorldMapper, https://www.worldmapper.org/, is a project developed by Danny Dorling and others, largely at the University of Sheffield in the United Kingdom. It now has nearly 700 maps in thirty general categories covering food, goods, income, poverty, housing, education, disease, violence, causes of death, and so forth. Their slogan is "mapping your world as you've never seen it before." It is well worth a visit, if not a journey.

Chapter 5

- Unlike most other classics in the history of data visualization, Playfair's main works—the *Atlas* and the *Statistical Breviary*—are available in a modern reprinting, edited and introduced by Howard Wainer and Ian Spence and published in 2005 by Cambridge University Press. A modern reader may be interested in reading Playfair's words to see how he faced the challenge of describing his novel charts to his audience around 1800. Equally well, one might be impressed with the quality

of the reprinting and the presentation of Playfair's plates, a number of which are on fold-out pages.
- The collection of research papers and biographical studies of Playfair by Ian Spence can be found at https://psych.utoronto.ca/users/spence/Research_WP.html.

Chapter 6

- Parts of the chapter are based on material presented in Friendly and Denis (2005).
- The remarkable capabilities of the ellipse and higher dimensional cousins (ellipsoids) are described in Friendly et al. (2013). Slides from a related talk can be found at https://datavis.ca/papers/EllipticalInsights-2x2.pdf.
- You can find more examples of spurious correlations on the web page of Tyler Vigen, https://www.tylervigen.com/spurious-correlations, and his accompanying book, *Spurious Correlations* (2015). Most of these result from plotting two different time series, using different scales for each on the *y*-axis, which now is usually considered a graphical sin, if not a high crime or misdemeanor.

Chapter 7

- This chapter draws heavily on Friendly (2008b), "The Golden Age of Statistical Graphics." This paper contains many more figures and greater depth on some of the topics discussed here.
- Friendly (2002) provides a review of Minard's graphical works. A complete bibliography with many images can be found at https://datavis.ca/gallery/minbib.html. A more recent work, Rendgen (2018), *The Minard System*, provides beautiful reproductions of all of Minard's statistical graphics, some of which were not well known before.
- Raymond Andrews has taken this further, with a visual catalog of Minard's works, with thumbnails, a timeline, and classification by content topic. See this at https://infowetrust.com/seeking-minard/.
- Some years ago, one of the authors issued a challenge for modern software designers to take Minard's data and either reproduce his graph

of the fate of Napoleon's Grand Army or take this graphic story further. A number of these are collected at http://www.datavis.ca/gallery/re-minard.php. A recent addition, an interactive chart by Norbert Landsteiner, https://www.masswerk.at/minard/, is one of the nicest interactive reproductions we have seen.

- Many of the developments in data-based graphics in this period stemmed from thematic cartography, the use of maps to present quantitative information in a geographical framework. The most complete discussion of the rise of graphical methods in this context in France in the nineteenth remains Gilles Palsky's 1996 book, *Des Chiffres et de Cartes*. Friendly and Palsky (2007) give a more extensive overview of the development of the use of maps and statistical diagrams to visualize nature and society.

Chapter 8

- The most comprehensive source on the development of thematic cartography is Arthur Robinson's *Early Thematic Mapping in the History of Cartography* (1982).
- Some historical connections among scientific discovery, visual explanation, and thematic maps and statistical diagrams are described in Friendly and Palsky (2007).
- The modern text by Slocum et al. (2008) covers thematic cartography and geographic visualization.
- Three-dimensional surfaces have long been studied as mathematical objects, described by equations, or (x, y, z) data and capable of being rendered realistically with lighting and shadows. See https://www.scratchapixel.com/lessons/3d-basic-rendering/rendering-3d-scene-overview for a readable introduction to this topic.

Chapter 9

- Maria Braun's *Picturing Time* (1992) is the most comprehensive treatment of the work of E.-J. Marey. It contains over 300 images of his mechanical devices, chronophotographs, and cinematic work.

- The Graphics Video Library of the Statistical Graphics Section of the American Statistical Association, http://stat-graphics.org/movies/, captures much of the history of dynamic graphics for data analysis over the past fifty years. It includes Kruskal's MDS video, Tukey's PRIM-9 video, and many more that illustrate and explain some of the important early developments in modern data visualization methods.
- Friedman and Stuetzle (2002) give a historical appreciation of John W. Tukey's work in the development of PRIM-9 and interactive graphics. Cook and Swayne (2007) provide some modern examples of interactive and dynamic graphics, using R and GGobi software.
- The collection of Hans Rosling's video presentations may be found at https://www.gapminder.org/videos-2/. One of the first, introducing the moving bubble chart, was a TED talk titled "The Best Stats You've Ever Seen," https://www.youtube.com/watch?v=hVimVzgtD6w. These are well worth watching.

Chapter 10

- Minard's graphic of the 1812 campaign has inspired many people to try to extend the graphic portrayal of this history. A collection of some of these and further background can be seen at http://www.datavis.ca/gallery/re-minard.php.
- Recently, an impressive interactive map and historical narrative of the Russian campaign with many illustrations was developed by the Russian News Agency, TASS, https://1812.tass.ru/, and was shortlisted for the Kantar Information Is Beautiful Awards.
- The entire collection of the graphic works of C. J. Minard is now available in a lovely volume curated by Sandra Rendgen (2018).
- A discussion of W. E. B. Du Bois's charts at the 1900 Paris Exposition Universelle and a wider selection of them can be found at https://hyperallergic.com/306559/w-e-b-du-boiss-modernist-data-visualizations-of-black-life/.
- The drafting of Plates 19 and 20 required more than standard graphics software. To learn the full story behind the development of these figures, see https://infowetrust.com/picturing-the-great-migration/.
- A related article is Andrews and Wainer (2017).

NOTES

REFERENCES

ACKNOWLEDGMENTS

INDEX

Notes

Introduction

1. GBD 2016 Alcohol Collaborators (2018), published online August 23, 2018. https://doi.org/10.1016/S0140-6736(18)31310-2.

2. A version of this quotation is attributed to the American businessman John Nasbitt.

3. We are unable to find a citation for this as a quotation, but this is the view that Tukey surely espoused. With his stature as a statistician, he revolutionized the field in recognizing data analysis as a broad discipline that includes statistics, in a similar way that the term *data science* is now being used to reorganize current practice and thinking. There are other phrasings of this idea connecting purpose to insight in other contexts. The noted computer scientist Richard Hamming stated that "The purpose of computing is insight, not numbers" (Hamming, 1962, Preface). Ben Shneiderman, a guru in data visualization, has often written that "The purpose of visualization is insight, not pictures," (e.g., in Card et al., 1999) a sentiment we heartily endorse.

4. Beniger and Robyn (1978).

5. Bertin (1973, 1977, 1983).

6. Tufte (1983, 1990, 1997, 2006).

7. The initial version can still be found at http://euclid.psych.yorku.ca/SCS/Gallery/milestone/.

8. At that time, the only comprehensive, book-length treatment of the history of data-based graphics was Howard Gray Funkhouser's [1898–1984] 137-page article in the journal *Osiris*, published in 1937. James R. Beniger, a sociologist specializing in communication, revived the study of this history around 1975 in a series of conference presentations. In 1978, Beniger and Robyn published an influential brief history of quantitative graphics in statistics that awakened modern interest in this topic. Other coherent accounts of graphical history tend to be focused more narrowly. One example of this is Arthur Robinson's 1982 magisterial reporting of the history of thematic mapping, which is justly famous among cartographers.

Another is Gilles Palsky's equally authoritative 1996 account of the development of thematic cartography in the nineteenth century; currently it is only available in French.

9. This approach is described in Friendly (2005, 2008a), but the essential idea was first expressed by Ernst Rubin in 1943. Some historians prefer the term *quantitative history*, and there are now a number of historical journals that focus on this topic. Even broader terms are cliodynamics and cliometrics (after Clio, the muse of history), which also treat history as a topic that can be studied quantitatively to ask how and why things changed over time.

10. See Friendly et al. (2015) for details and examples of these methods.

1. In the Beginning . . .

1. Linguists more carefully distinguish among ideographic scripts (in which graphic symbols represent concepts or ideas), pictographic scripts (in which the symbols are iconic pictures), and logographic scripts (where the glyphs represent sounds or morphemes). Egyptian hieroglyphs were initially thought to be ideographic, but they actually are logographic, as they use visual symbols to represent sounds and syllables of spoken language.

2. Bachi (1968).

3. Tukey (1977, p. 17).

4. Henri Breuil was a French Jesuit priest, archaeologist, anthropologist, ethnologist, and geologist. He is noted for his studies of cave art in the Somme and Dordogne valleys as well as in Spain, Portugal, Italy, Ireland, China, and elsewhere.

5. Harari (2015).

6. See https://news.bbc.co.uk/2/hi/science/nature/871930.stm.

7. Known as Ulysses in Roman myths.

8. The history of cartography has many sources, but among these, the project of the same name at the University of Wisconsin (https://geography.wisc.edu/histcart/) is notable. There are now six volumes, some published online and in print form by the University of Chicago Press (https://www.press.uchicago.edu/books/HOC/index.html).

9. See Hoff and Geddes (1962) for the history of graphic recording.

10. Yet naked empiricism and the mere transcribing of numbers into pictures did not meet with universal approval. As late as 1847, Luke Howard [1772–1864], sometimes called the "father of meteorology," apologized for his methodology and referred to it as an "autograph of the curve . . . confessedly adapted rather to the use of the dilettanti in natural philosophy as that of regular students" (Howard, *Barometrographia*, 1847).

11. Plot's proposed method of crowd-sourcing weather data and his assessment of its potential value would later bear great fruit in Francis Galton's 1863 spectacular

discovery of weather patterns in the northern hemisphere, as we describe in Chapter 7.

12. This practical use spurred mathematicians like Edmund Halley (1693) and Abraham de Moivre (1725) to develop methods to calculate life expectancies from such data.

13. Biderman (1978).

14. Interestingly, one of Playfair's earliest emulators was the banker Samuel Tertius Galton (the father of Francis Galton, and hence the biological grandfather of modern statistics), who, in 1813, published a multiline time-series chart of the money in circulation, rates of foreign exchange, and prices of bullion and wheat (S. T. Galton, 1813). Ironically, had Galton paid close enough attention to his own graphs he might have been able to foresee the financial crisis of 1831, which created a ruinous run on his own bank. See http://www.danielebesomi.ch/research/diagrams.

15. For more about the remarkable life and accomplishments of William Playfair (including the fascinating story of his attempted blackmail of Lord Archibald Douglas), the interested reader is referred to Spence and Wainer (1997, 2005), Wainer (1996), and Wainer and Spence (1997).

16. Hacking (1990).

17. This enthusiasm was far from universal. In Britain, influential statisticians proudly called themselves "statists," meaning that they viewed their role as compilers of "facts," which were typically presented as tables rather than graphs.

2. The First Graph Got It Right

1. So many discoveries and inventions are named for someone other than the original inventor, that this phenomenon in the history of science is now called "Stigler's Law of Eponomy," after Stephen Stigler (1980), who, illustrating this law self-referentially, attributes the idea to Robert Merton, an influential sociologist of science.

2. Funkhouser (1937, p. 277).

3. A ducat was a gold coin containing about 3.545 grams of gold; 6,000 ducats was then equal to 21.27 kilograms or 47.6 pounds of gold. This was a substantial amount of money then, as it is now, roughly $2 million.

4. Stigler (2016, ch. 1).

5. Stigler (1986).

6. Tufte (1997).

7. See https://www.mathpages.com/home/kmath151/kmath151.htm for this and other examples of the use of coded descriptions of scientific discoveries.

8. Credit for the careful archival work leading to these discoveries is due to Gustavo Vieira.

3. The Birth of Data

1. His explanation was that males, who had to labor to seek food, perhaps with danger, were more subject to accidental death. Therefore, the wise Creator brought forth more males than females.

2. Campbell (2001) describes some of the limitations in Arbuthnot's data and logic.

3. See Hacking (1990, pp. 64–66).

4. Hacking (1990, Chapter 9).

5. See Friendly and Palsky (2007) and Robinson (1982) for comprehensive histories. Delaney (2012) presents a nice collection of landmark thematic maps in a variety of disciplines.

6. Robinson (1982, p. 55).

7. A high-resolution zoomable image can be found at https://digital.library.cornell.edu/catalog/ss:19343162.

8. Tufte (1983, 1990).

9. Tufte (1990, p. 67).

10. See Koch (2000) for a history of disease maps and Palsky (1996) for the development of thematic cartography.

11. Guerry had written to Quetelet in early 1831 describing his initial findings and conclusions. Quetelet later quoted parts of Guerry's letters, but claimed they merely supported his own arguments. But André-Michel Guerry was a modest man, both by birth and personality. He could have publicly defended his rights to the discovery of the regularity in crime statistics against the well-connected Quetelet, who claimed sole honors for himself. Quetelet, an eminent astronomer and mathematician, had a larger, bolder vision and was also a tireless self-promoter. Guerry, a retiring young lawyer and amateur statistician, was content to simply continue his labors. Sir Leon Radzinowicz put it thus: "Quetelet was like the huge tree that tends to dwarf its neighbors. The qualities of the two were, indeed, complementary, and in substance their contributions were virtually parallel. Thus, it may be fairly asserted that the sociology of crime owes its inception to Guerry as surely as it does to Quetelet" (Radzinowicz, 1965, p. 1048).

12. Émile Durkheim (1897) adopted this approach to the study of suicide, but without much credit to Guerry or other moral statisticians.

13. Diard (1866).

4. Vital Statistics

1. The Births and Deaths Registration Act, 1836, https://tinyurl.com/jykzwgt.

2. Published in J. R. McCulloch, ed., *A Statistical Account of the British Empire* (London, 1937), 2 vols., pp. 567–601.

3. W. Farr, Letter to the Registrar General, in *First Annual Report of the Registrar General* (London: HMSO, 1839). See also M. Whitehead (2000), https://www.who.int/bulletin/archives/78(1)86.pdf, on which this account draws.

4. See J. N. Hayes (2005), pp. 214–219.

5. Review, *Lancet*, 1852, 1, 268. See also John Eyler (1973), on which some of this discussion relies.

6. Tufte (1983, ch. 8).

7. The first known appearance is a little-known paper, Guerry (1829), in the *Annales d'hygiène publique et de médecine légale*, showing such diagrams of weather and physiological phenomena.

8. From General Register Office (1852, pp. clii–clvii). The data are contained in the HistData package for R, as the data set Cholera.

9. In the language of modern statistical software (R), this would be expressed as `cholera ~ water * elevation`, where `water` is a classification factor.

10. A modern analysis of these data would not use simple linear regression or classical linear models for the rate of cholera mortality per 10,000 population. More appropriate would be generalized linear models, such as logistic regression for the number of deaths relative to population or Poisson regression models.

11. Snow (1849b), "On the pathology and mode of transmission of cholera."

12. Snow (1849a), expanded and republished in Snow (1855).

13. General Register Office (1852, pp. lxxvi–lxxvii).

14. J. Jameson, *Report on Cholera in Bengal*, cited by Farr (1852), p. lxxvi.

15. See https://www.hetwebsite.net/het/schools/rss.htm for a description of the early history of the Statistical Society of London. William Farr was elected a fellow of the SSL in 1839 and then president in 1871–1873; he gave two presidential addresses in which he recounts some important milestones in this history.

16. Snow's book can be found online at https://www.ph.ucla.edu/epi/snow/snowbook.html.

17. For a thorough discussion of the roles of Farr and Snow, see Eyler (2001).

18. Koch (2000, 2004, 2011).

19. Tufte (1997, pp. 27–37).

20. That honor is usually accorded to the physician Valentine Seaman, who mapped an outbreak of yellow fever in New York City in 1789. See Wallis and Robinson (1987) for the initial discussion in the context of thematic mapping and Koch (2011) for a more extended treatment.

21. Tufte (1997, p. 30).

22. *Report on the Cholera Outbreak in the Parish of St. James, Westminster: During the Autumn of 1854* (London: J. Churchill).

23. Snow (1855, p. 109).

24. Snow (1855, p. 109).

25. See Koch (2004) for a survey of some of these attempted enhancements.

26. Gilbert (1958).

27. These are contained in the HistData package for R. They were originally digitized in 1992 by Rusty Dodson at the National Center for Geographic Information and Analysis of UC Santa Barbara and made public by Waldo Tobler. Examples in the package allow you to reproduce Snow's map in a variety of ways.

28. The intention of the Voronoi polygons is to define the most likely (closest) pump from which people would draw their water. They are easy to compute because they use Euclidean distance. The main drawback is that roads and actual walking distance play no role in the choice of pump: the method assumes that people can walk through walls to get to their preferred pump.

29. We are grateful to Peter Li for assistance with this. His cholera package for R provides other versions of Snow's data sets and facilities for more detailed analysis of the pump neighborhoods that Snow tried to illustrate in Figure 4.8.

30. There are numerous other reproductions of Snow's map, using modern GIS or statistical graphics software. See Shiode et al. (2015) for some examples using GIS. Among interactive versions, one of our favorites was developed by ArcGIS, https://www.arcgis.com/apps/PublicInformation/index.html?appid=d7deb67f810d46dfacb80ff80ac224e9.

31. Koch (2013).

32. Edmund Parkes (1855), an established medical authority, reviewed Snow's *On the Mode of Communication of Cholera* in the year of the publication of its second edition. Among many other complaints, he chided Snow for his use of anecdotal accounts of the well from which a victim may have drawn his water compared with that of a neighboring survivor, claiming such evidence to be worthless.

33. This phrase is attributed to Edward T. Cook's 1913 biography, *The Life of Florence Nightingale*, available at https://archive.org/details/lifeflorencenigh01cook. Her biography as a statistician is told by Kopf (1916).

34. Kopf (1916).

35. One result of this would later be called "mobile army surgical hospitals," dedicated to saving as many lives as possible, under battlefield conditions. The acronym MASH arose from this term and became the title of a movie and TV series about combat medicine during the Korean War.

36. Nightingale (1858).

37. Nightingale (1858, Diagram K, p. 47).

38. Nightingale (1858, p. 298).

39. Stigler (2016, ch. 7).

40. This graph was suggested by Gelman and Unwin (2013), in a thoughtful discussion of the differences between "Infovis" and statistical graphics.

5. The Big Bang

1. This chapter owes a great deal to the work of Ian Spence. Segments of it come from work that was jointly authored by Spence and Wainer but originated with Spence. We are grateful for his blessing to use that work here.

2. Today, this practice is often deprecated, following Tufte's 1983 call for minimizing "non-data ink" in charts, but the general idea of showing reference indicators in charts still stands.

3. Venn (1880).

4. Cleveland and McGill (1984a).

5. A survey by Toronto psychologist Ian Spence in 1998 found that 10 percent of all charts in the popular press were pie charts. They were much rarer in scientific and business publications.

6. France is admirably designed for use in thematic maps because the departments are all of comparable size; this is no accident. Napoleon decreed that their boundaries be set so that any citizen could ride by horse to the administrative center within one day.

7. Tukey (1977, p. vi).

8. He is downright insulting in his chapter on Turkey, where he says, "One of the finest and fairest portions of Europe, where arts, science, and literature once flourished in a high degree, is now possessed by the most ignorant, indolent, and debased race of men that ever encumbered the face of the globe" (p. 53).

9. Stamp (1929, pp. 258–259). The actual source of the quote is Harold Cox, an English judge and member of Parliament.

10. This plot reproduces one prepared by William Cleveland (1994, p. 228). The data are recorded in the data set `East Indies Trade` in the R package GDAdata.

11. Daniel Rosenberg and Anthony Grafton (2010) provide a compelling history of the development of the timeline as a tool for visualizing history in their *Cartographies of Time.*

12. See https://en.wikipedia.org/wiki/A_New_Chart_of_History.

13. It is unlikely that Playfair had any real data on this, and never in the text does he say what the vertical axis of each mini-chart represents. It is most likely that he is reporting his impressions from a close and imaginative reading of Edward Gibbon's *Decline and Fall of the Roman Empire* and Adam Smith's *Wealth of Nations.*

14. The data come from https://voteview.com, a website that allows users to view every congressional roll-call vote in American history on a spatial map of political ideology. The DW-NOMINATE scores were developed by political scientist Kenneth Poole (2005) and others in the 1980s as a scaling method to locate points in "political space" based on the similarity of the voting records of legislators. The analysis summarized in Plate 9 was based on 12,000 legislator-year records and over 100,000 roll-call votes.

15. John Stuart Mill argues this most forcefully: "It is a piece of idle sentimentality that truth, merely as truth, has any inherent power denied to error of prevailing against the dungeon and the stake. . . . The real advantage which truth has consists in this, that when an opinion is true, it may be extinguished once, twice or many times, but in the course of ages there will generally be found persons to rediscover it, until some one of its reappearances fall on a time when from favorable circumstances it escapes persecution until it has made such headway as to withstand all subsequent attempts to suppress it" (John Stuart Mill, *On Liberty* [1859], ch. 2).

16. W. Playfair (1822–1823), unpublished ms. held by John Lawrence Playfair, Toronto, Canada, transcribed and annotated by Ian Spence.

17. Hankins and Silverman (1999, p. 120).

18. Indeed, in discussing his "lineal arithmetic" in the *Commercial and Political Atlas*, Playfair devoted six pages to explaining what was represented in a graph. He likened the height of a bar or line showing money to that of a pile of coins. Putting these side by side for several months or years, the heights would be proportional to the actual amount of money.

19. For a discussion of Jevons as a statistician, and some of his development of statistical graphics, see Stigler (1999, 66–79).

20. Marshall (1885).

21. Levasseur (1885).

22. Spence et al. (2017).

6. The Origin and Development of the Scatterplot

1. Tufte (1983) estimated that 70–80 percent of graphs used in modern scientific publications are scatterplots; see also Cleveland and McGill (1984b) for some modern enhancements.

2. We are grateful to Stephen Stigler for pointing out the connection of Cotes's work with the history of the scatterplot. His 1986 book *The History of Statistics* is the authoritative source on the development of the measurement of uncertainty in the history of statistics.

3. Had Cotes wanted to illustrate the center of gravity visually, he would have made the *sizes* of the points p, q, r, and s proportional to their weights, so the solution Z could be *seen* to be their center of gravity. As far as we know, the first use of this idea in a data graphic was by C. J. Minard in a cartograph of Paris designed to answer the question of where to build a new post office. Minard's visual solution was the center of gravity of the populations of the arrondissements, which he showed by proportional squares. See figure 6 in Friendly (2008b).

4. See Bullynck (2008) for a thorough discussion of Lambert's work and his philosophy of the graphical method.

5. Lambert (1765, pp. 430–431).

6. Playfair (1821, p. 31).

7. With good reason, because the idea of relating one time series to another by ratios (or index numbers, such as are now used to show economic data in "constant dollars") would not occur for another half-century, in the work of Jevons (1863).

8. Stigler (1986).

9. Herschel (1833b, p. 199).

10. Herschel (1833b, p. 171).

11. Herschel (1833b, para. 14, p. 178).

12. Herschel (1833b, sect. 2, pp. 188–196).

13. Herschel (1833a, p. 35).

14. Herschel's observational data was transcribed from his table on p. 35 of Herschel (1833a). The data he calculated from his smoothed curve were transcribed from his Table 1 on p. 190 of Herschel (1833b). These data sets are available as `Virginis` in the HistData package for R.

15. This discussion has been informed by Hankins (2006), who gives a far more detailed and nuanced discussion of Herschel's methods and results.

16. Hilts (1975, fig. 5, p. 26).

17. Galton (1886, p. 254) describes his process quite explicitly: "I will call attention to the form in which the table of data [Table I] was drawn up. . . . It is deduced from a large sheet on which I entered every child's height opposite to its mid-parental height, and in every case each was entered to the nearest tenth of an inch. Then I counted the number of entries in each square inch and copied them out as they appear in the table."

18. In Galton's earliest work, he used the median y for each class of x values (Stigler, 1986), perhaps because it was easier to calculate. This idea would later reappear in various guises (resistant lines, robust estimation).

19. Galton selected sweet peas for this experiment because, as he later described in *Natural Inheritance*, "they do not cross-fertilize; . . . they are hardy, prolific, of a convenient size to handle, and nearly spherical; their weight does not alter perceptibly when the air changes from damp to dry" (Galton, 1889, p. 80).

20. This experiment was also noteworthy because he carefully considered how best to measure the "size" of the parent and child seeds: by their average weight or average diameter? In *Natural Inheritance*, he gives a table showing the weight of one seed in a packet, in grains, and the corresponding measurement of the length of 100 seeds in a row, translated to the average diameter of one seed in hundredths of an inch.

21. The data set used here was reported by in a tutorial article by Stanton (2001), and then incorporated into the R psych package as the data set `peas`.

22. This was another crowd-sourced study. Galton sent notices to newspapers, asking parents to record information about themselves and their children and mail it to him. He carefully noted that wives on average were slightly shorter than their husbands, so his measure of mid-parent height multiplied the heights of mothers by 1.08 before averaging them.

23. Friendly et al. (2013).

24. Pearson (1901).

25. The story of Galton's invention of the concept of correlation is best told by Stephen Stigler, in his book *The History of Statistics* (Stigler, 1986, chs. 8–9). In a short article (Stigler, 1989), he relates the full story behind this invention.

26. Online at https://galton.org/essays/1890-1899/galton-1890-nareview-kinship-and-correlation.html.

27. Spence and Garrison (1993).

28. This data set is available as `starsCYG` in the **robustbase R** package.

29. The Phillips curve has the three-parameter exponential form, $y + a = b x^c$, or the (nearly) linearized form $\log(y + a) = \log b + c \log x$, where y is the change in wage rates and x is unemployment. Phillips could have fit this function to *all* the data using available methods for curve fitting. Instead, he apparently applied some degree of hand-eye-brain smoothing (as Herschel had done), for he chose six representative points, shown by the crosses in Figure 6.19, and fit the curve to them, using least squares to estimate b and c and trial and error to estimate the offset a.

30. Out of seventeen pages, a total of six are devoted to scatterplots, a rather large 35 percent of journal space.

31. The earliest eponymous econometric curves we have found are from Engel (1857); they describe the distribution of the percentage of an individual's expenditures on a given commodity (e.g., food or housing) in relation to total income.

32. It was inspired by Alberto Cairo, *The Functional Art* (2012).

33. Presumably with tongue planted firmly in cheek, as is the rest of the discussion of this topic.

34. https://tinyurl.com/8bbweav.

35. The scatterplot matrix originated with Hartigan (1975). It has subsequently been extended in many ways, including pairwise "mosaic" plots of categorical variables (Friendly, 1999), and generalized pairs plots (Emerson et al., 2013) show a mixture of quantitative and categorical variables in various formats.

7. The Golden Age of Statistical Graphics

1. Friendly (2005), Friendly et al. (2015).

2. See Friendly (2008a) for a description of these historical periods.

3. Funkhouser (1937), Palsky (1996).

4. Arthur Robinson (1982, p. 57) credits the influence of lithography in this time as important as that of rapid copying techniques starting with Xerox machines in our time.

5. The most extensive collection of such devices we know is in the Conservatoire des Artes et Métiers in Paris. Among these are several versions of Blaise Pascal's

pascaline, constructed with spoked metal dials and with a mechanism to carry an increment to the next higher digit. A visit there is well worth a detour.

6. See Friendly and de Saint Agathe (2012) for the story of the loss of this instrument, and its rediscovery of sorts in the archives of the Conservatoire des Artes et Métiers.

7. This image is available at https://www.datavis.ca/gallery/images/Lalanne.jpg.

8. These are reproduced in the marvelous book *The Minard System* by Sandra Rendgen (2018).

9. Tufte (1983, p. 40).

10. The battle of Borodino, shown about midway in Minard's graph, was a major turning point in the French invasion. At this point, Tchaikovsky's score calls for five Russian cannon shots to counter a fragment of *La Marseillaise*, the French national anthem.

11. Stephen Stigler, personal communication, quoted by Wainer (2005).

12. We are indebted to Stephen M. Stigler for providing high-resolution scans of Galton's figures from his personal copy of *Meterographica*.

13. Galton (1866).

14. It is hard to resist the comment that Galton showed that you *do* need a weatherman to show which way the wind is blowing. See Monmonier (1999) for a cartographic perspective on the history of weather maps.

15. Although the statistical albums of different countries were separate works, it is worth noting that the central characters in France (Émile Levasseur, Émile Cheysson), Germany (Georg von Mayr, Hermann Schwabe), England (William Farr, Joseph Fletcher), the United States (Francis Walker, Henry Gannett), and elsewhere were well acquainted with each others' work, from international expositions, conferences, and informal exchanges. Among these, the International Statistical Congresses, organized in 1857 by Quetelet and others, soon became an international forum for debating the use of graphical methods and for attempts to develop international standards; see, for example, International Statistical Congress (1858), Schwabe (1872), Palsky (1999).

16. Chevallier (1871, p. 17); see also the account by Raymond J. Andrews, https://infowetrust.com/seeking-minard, which contains an animated graphic of the portrait of Minister Eugène Rouher from the Musée Mandet in Riom, France.

17. Faure (1918, p. 294); see also the discussion in Palsky (1996, pp. 141–142).

18. The entire collection of the *Albums de Statistique Graphique* was first acquired by the authors and other friends around 1998. Recently, David Rumsey acquired another set and has put digitized copies online. See https://www.davidrumsey.com/luna/servlet/s/nl72bu for the complete collection.

19. This name, which originated with Becker et al. (1996), evokes a grid-like arrangement of plots. Other software calls these "lattice plots" or "faceted displays."

20. For example, the time to travel from Paris to Toulouse in 1650 was 330 hours by horseback; with increasing development of railways, this decreased to 104 hours by 1814 and to only 15.1 hours by 1887. Today the same journey takes about 5 hours by the TGV high-speed rail route through Bordeaux. Montpellier and Marseilles, approximately the same geographical distance from Paris, became progressively closer in travel time. Today, the TGV takes you to Marseilles in about 3.25 hours.

21. The WorldMapper project, http://worldmapper.org, provides a huge collection of cartograms of health, social, and economic topics. In political and election reporting, cartograms are now preferred for showing the distribution of votes in ways not distorted by the visual impression of area.

22. Copies of these census atlases on CD, containing high-resolution, zoomable images, may be obtained from the Historic Print and Map Co., https://www.ushistoricalarchive.com.

23. From Ex. Doc. No. 9 of the 42nd Congress, 1871–1873 (Walker, 1874, p. 1).

24. See Hofmann (2007) for a detailed re-analysis of this chart from a modern perspective.

25. Dahmann (2001).

26. Statistischen Bureau (1897, 1914). These are available on CD, in a form that integrates the explanatory text with the plates, from the Swiss Bundesamt für Statistik, order number 760-0600-01.

27. Dahmann (2001).

28. This term was introduced by Friendly and Denis (2000).

29. Of course, Galton, Pearson, and even Fisher were enthusiastic graph people, but many of those who extended their ideas (e.g., F. Y. Edgeworth, G. U. Yule, and even Pearson himself) turned their attention to mathematical and analytic aspects of correlation, regression, and general theories of statistical distributions. Later developments of likelihood-based inference, decision theory, and measure theory served to increase the sway of more formal mathematical statistics. "Indeed, for many years there was a contagious *snobbery* against so unpopular, vulgar and elementary a topic as graphics among academic statisticians and their students" (Kruskal, 1978, p. 144; emphasis in the original).

30. Historians of science might eschew our use of the designation "Modern dark ages," and instead refer to this as merely a progression in what Thomas Kuhn (1970) called a period of normal science following some scientific revolution. This is fine, as far as it goes, but misses the main point here: the translation of theory into popular practice and understanding.

8. Escaping Flatland

1. Similarly, Delaney (2012) describes the development of thematic maps in his title as *First X, Then Y, Now Z,* indicating that the simplest idea was that of place

names (*X* marks the spot), followed by (*X*, *Y*) coordinates of latitude and longitude, and finally using coloring, shading, and symbols as a thematic *Z* layer.

2. Cassini de Thury, sometimes called Cassini III, was the son of the famous astronomer Jacques Cassini, who compiled the first tables of the orbital motions of Saturn's satellites. Jacques, in turn, was the son of astronomer Gian Domenico Cassini, who in 1718 completed the measurement of the arc of the meridian (longitude line) between Dunkerque and Perpignan to determine the shape of the earth. All three Cassinis were directors of the Paris Observatory.

3. An earlier example of a map showing geomagnetic lines was published in 1602 by the French astronomer and geographer, Guillaume de Nautonier of Castelfranc [1557–1620], also for the purpose of contributing to the determination of longitude. It was not, however, derived from hard data. The title reads *The whole world described from expertly compiled sources. . . .* Mandea and Mayaud (2004) give a historical appreciation. Nautonier built one of the first astronomical observatories in Europe, which still stands on the grounds of his estate near the town of Montredon-Labessonnié, in the Tarn, France.

4. Such was Halley's scientific stature and the importance of this project that the Royal Navy gave him command of the *Paramour*, even though his previous experience at sea had only been as a passenger. See Thrower (1981) for a complete account of Halley's voyages.

5. Murray and Bellhouse (2017).

6. The caption reads: "Isochronic Passage Chart for Travellers showing the shortest number of days journey from London by the quickest through routes and using such further conveyances as are available without unreasonable cost. It is supposed that local preparations have been made and that other circumstances are favorable."

7. We used this technique in the enhancement of John Snow's cholera map (see Plate 3).

8. Lalanne (1879).

9. The method was reported to the French Academy of Sciences on February 17, 1845. The version shown here appeared in an appendix in a course on meteorology edited by L. F. Kaemtz.

10. Some spectacular examples of current software for realistic 3D maps and surfaces use the rayshader package for R, https://www.rayshader.com/.

11. There were no detailed population data for the years prior to 1751. Swedish population statistics were centralized around 1860, which led to a substantial improvement from 1861 onward. In the period before 1860 there were problems with the age distribution in the censuses, later corrected by Gustav Sundbärg in 1908. Perozzo did not have this corrected data available when drawing his graph. See the report by Dana Glei and others at https://bit.ly/3c7yq0y.

12. This is a "best guess" conjecture by demographers Tim Riffe and Sebastian Kluesner (personal communications, Sept. 2017).

13. Perozzo (1881).

14. See Brian (2001) for a discussion of the connections among scientific culture, theories of probability, and surrealism in the later part of the nineteenth century.

15. Lexis (1875).

9. Visualizing Time and Space

1. Principal components analysis, factor analysis, multidimensional scaling, and correspondence analysis are just a few of these methods.

2. The exact strength of gravity actually varies with location on the Earth, north-south latitude and altitude being the primary sources of deviation from a constant value, but even local topography and geology can cause anomalies. A precise mapping of gravity variation and anomalies was begun in 2002 by NASA's GRACE (Gravity Recovery and Climate Experiment) satellite mission. We mention this because, as befits this chapter, NASA created a lovely animated visualization of a lumpy Earth, showing the deviation from standard gravity by color and shape. See this image at https://en.wikipedia.org/wiki/File:GRACE_globe_animation.gif.

3. Antony Tudor [1909-1987] at the American Ballet Theater developed this idea and used the phrase "let the movement tell the story." He was known as the psychological choreographer.

4. Braun (1992, p. 45).

5. Muybridge (1887), *Human and Animal Locomotion*, plate 626, shows a rider on the horse "Annie G.," galloping. You can see this in a modern animation that also tracks the motion of the horse's feet at https://bit.ly/3sYovQV.

6. An equivalent view today is an animated GIF image constructed from these frames shown sequentially and repeating (see https://tinyurl.com/horse-in-motion-gif).

7. Marey, *La machine animale* (1873; translated as *Animal Mechanism*, 1884, pp. 6–7).

8. Anonymous (1869).

9. *Falling Cat*, https://www.imdb.com/title/tt2049440/.

10. Marey (1894).

11. Anonymous (1894).

12. Kane and Scher (1969).

13. From a web page by Mike Williams, 2004, "Realistic shapes," which is now defunct.

14. As recently as 1989, one of us (MF) began to write a book (Friendly, 1991) on statistical graphics, with hundreds of graphs produced on a four-color Calcomp pen plotter. Each final figure went through very many versions to get it right, and

the work wore out several plotters, not to mention the patience of the computer operators who maintained them.

15. See http://bit.ly/3a6Oahv.

16. Available at https://www.youtube.com/watch?v=p_0WXK8wupU.

17. Ekman (1954). The data was based on the average rating by thirty-one subjects with normal color vision where each of the ninety-one pairs of colors was rated on a 5-point scale (0 = no similarity up to 4 = identical).

18. If you are wondering about the large gap between the ends of this color spectrum, the color purple would normally be there but, unlike the distinct wavelengths at the blue and red ends, purple is a composite color, made by combining red and blue light.

19. The early history is described by Roger Eckhardt, "Stan Ulam, John von Neumann, and the Monte Carlo Method," *Los Alamos Science* (1987), Special Issue, https://bit.ly/3ooLnFv.

20. Certain kinds of electronic noise can be used to produce sequences of random numbers, but these sequences can never be duplicated exactly. A digital computer can generate deterministic sequences from an algorithm, but they cannot be truly random. The challenge is to construct deterministic sequences that are close enough to random.

21. Marsaglia (1968).

22. Fisherkeller et al. (1974).

23. You can watch this at https://www.youtube.com/watch?v=B7XoW2qiFUA.

24. Tukey (1962).

25. The story of this Nobel Prize is well described at https://www.nobelprize.org/educational/medicine/insulin/discovery-insulin.html.

26. Reaven and Miller (1979, p. 22).

27. The first idea of the graphical user interface was proposed by Ivan Sutherland in his 1963 PhD thesis at MIT (Sutherland, 1963). An extensive history of the GUI is given in https://en.wikipedia.org/wiki/History_of_the_graphical_user_interface; Jeremy Reimer, https://arstechnica.com/features/2005/05/gui/, provides a more discursive history focusing on the development of ideas and techniques.

28. Newton (1978), McDonald (1982), Buja et al. (1988).

29. Donoho et al. (1988).

30. Velleman and Velleman (1985).

31. Friedman and Tukey (1974).

32. Asimov (1985).

33. The term was coined in Tukey and Tukey (1985) to refer to scatterplot diagnostics.

34. Rosling and Johansson (2009).

35. See https://www.gapminder.org/videos/200-years-that-changed-the-world/.

10. Graphs as Poetry

1. Mumford (2000, p. 12).

2. Ignat Solzhenitsyn, April 10, 2004, from his introduction to an all-Mozart concert he conducted at Verizon Hall, Philadelphia, PA.

3. See Friendly (2002) for a timeline of Minard's works.

4. We discussed the wonderful falling cat films of Étienne-Jules Marey in Chapter 9. He also is well known for designing the easy-to-use graphical schedule for all trains between Paris and Lyon shown in Figure 1.7.

5. The details here come from the Greek historian Polybius [c. 200–circa 118 BC], *The Histories*, on which Minard and later scholars have relied.

6. Minard's legacy and personal history are an active topic among historians of data visualization. The complete collection of Minard's graphic works were recently published by Rendgen (2018). A group of our colleagues recently discovered his last Paris address and his tomb in Montparnasse Cemetery. This topic will be the subject of a follow-up article.

7. Chevallier (1871). An English translation was prepared by Dawn Finley, https://www.edwardtufte.com/tufte/minard-obit. For further detail on Minard's life, see Friendly (2002) and a biography, http://datavis.ca/gallery/minard/biography.pdf.

8. Geneen (1985, ch. 9, p. 151).

9. Du Bois's account of this exhibit appears in Du Bois (1900). Nearly the entire collection, comprising over 200 photographs and numerous charts and maps, has been digitized by the Library of Congress. See http://bit.ly/369Oxa7.

10. Although all slaves in the Confederate states were freed by Lincoln's proclamation in 1863, it was not until the ratification of the Thirteenth Amendment on December 6, 1865, that all slavery was truly abolished in the United States.

11. It is hard to resist a critique of Du Bois's choice of colors in this thematic map. He is trying to show the relative numbers of Negroes in the various states, expressed as "Negroes to the square mile," indicated in the legend. He uses yellow, blue, red, brown, and black to represent, respectively, less than 1, 1–4, 4–8, 8–15, 15–25. But the visual impression of the categories greater than 1 is muddled. The 1–4 category in blue dominates the display; the 4–8 category in red sharply contrasts with this, and the highest category, 15–25 in black, gets lost among the blue and brown (8–15). To be fair to Du Bois, color was not widely used in data graphics at this time, except to indicate discrete, unordered categories. An understanding of color scales for ordered categories of amounts only developed later.

12. Lemann (1991).

13. Previously the census included slaves as merely counts within various categories (e.g., age and sex) as part of the household goods of their owners.

14. This section borrows from Andrews and Wainer (2017).

15. Minard (1862a). You can see this at https://bit.ly/3caPS47.

16. Henry D. Hubbard in the Preface to Brinton (1939).

References

A few items in this reference list are identified by shelfmarks or call numbers in the following libraries:

- **BeNL**: Belgian National Library;
- **BL**: British Library, London;
- **BNF**: Bibliothèque Nationale de France, Paris (Tolbiac);
- **ENPC**: École Nationale des Ponts et Chaussées, Paris;
- **LC**: Library of Congress;
- **UCL**: University College London.

Achenwall, G. (1749). *Staatsverfassung der heutigen vornehmsten europäischen Reiche und Völker im Grundrisse.* N.p.

Andrews, R. J., and Wainer, H. (2017). The Great Migration: A graphics novel featuring the contributions of W. E. B. Du Bois and C. J. Minard. *Significance*, 14(3), 14–19.

Anonymous. (1869). The velocity of insects' wings during flight. *Scientific American*, n.v. (16), 241–256.

Anonymous. (1894). Photographs of a tumbling cat. *Nature*, 51, 80–81.

Arbuthnot, J. (1710–1712). An argument for divine providence, taken from the constant regularity observ'd in the births of both sexes. *Philosophical Transactions*, 27, 186–190.

Asimov, D. (1985). Grand tour. *SIAM Journal of Scientific and Statistical Computing*, 6(1), 128–143.

Bachi, R. (1968). *Graphical Rational Patterns: A New Approach to Graphical Presentation of Statistics*. Jerusalem: Israel Universities Press.

Bauer, C. P. (2017). *Unsolved!: The History and Mystery of the World's Greatest Ciphers from Ancient Egypt to Online Secret Societies*. Princeton, NJ: Princeton University Press.

Becker, R. A., Cleveland, W. S., and Shyu, M.-J. (1996). The visual design and control of trellis display. *Journal of Computational and Graphical Statistics*, 5(2), 123–155.

Beniger, J. R., and Robyn, D. L. (1978). Quantitative graphics in statistics: A brief history. *American Statistician*, 32, 1–11.

Bertin, J. (1973). *Sémiologie graphique*. 2nd ed. The Hague: Mouton-Gautier. Trans. William Berg and Howard Wainer, published as *Semiology of Graphics*. Madison: University of Wisconsin Press, 1983.

Bertin, J. (1977). *La Graphique et le traitement graphique de l'information*. Paris: Flammarion. Trans. William Berg, Paul Scott, and Howard Wainer, published as *Graphics and the Graphical Analysis of Data*. Berlin: De Gruyter, 1980.

Bertin, J. (1983). *Semiology of Graphics*. Trans. W. Berg. Madison: University of Wisconsin Press.

Biderman, A. D. (1978). Intellectual impediments to the development and diffusion of statistical graphics, 1637–1980. In *Proceedings of the First General Conference on Social Graphics*. Leesburg, VA.

Braun, H., and Wainer, H. (2004). Numbers and the remembrance of things past. *Chance*, 17(1), 44–48.

Braun, M. (1992). *Picturing Time: The Work of Etienne-Jules Marey (1830–1904)*. Chicago: University of Chicago Press.

Brian, E. (2001). Les objets de la chose. Théorie du hasard et surréalisme au xx siècle. *Revue de Synthèse*, 122, 473–502.

Brinton, W. C. (1939). *Graphic Presentation*. New York: Brinton Associates.

Buja, A., Asimov, D., Hurley, C., and McDonald, J. A. (1988). Elements of a viewing pipeline for data analysis. In W. S. Cleveland and M. E. McGill, eds., *Dynamic Graphics for Statistics*. Pacific Grove, CA: Brooks / Cole.

Bullynck, M. (2008). Presentation of J. H. Lambert's text, "Vorstellung der gröÿen durch figuren." *Electronic Journal for History of Probability and Statistics*, 4(2), 1–18.

Cairo, A. (2012). *The Functional Art: An Introduction to Information Graphics and Visualization*. 1st ed. Thousand Oaks, CA: New Riders Publishing.

Campbell, R. B. (2001). John Graunt, John Arbuthnott, and the human sex ratio. *Human Biology*, 73(4), 605–610.

Chevalier, L. (1958). *Classes laborieuses et classes dangereuses à Paris pendant la première moitié du XIXe siècle*. Paris: Plon. Translated by F. Jellinek. New York: H. Fertig, 1973.

Chevallier, V. (1871). Notice nécrologique sur M. Minard, inspecteur génèral des ponts et chaussées, en retraite. *Annales des ponts et chaussées*, 5(2), 1–22. Trans. Dawn Finley. https://www.edwardtufte.com/tufte/minard-obit.

Cleveland, W. S. (1994). *The Elements of Graphing Data*. Summit, NJ: Hobart Press.

Cleveland, W. S., and McGill, R. (1984a). Graphical perception: Theory, experimentation and application to the development of graphical methods. *Journal of the American Statistical Association*, 79, 531–554.

Cleveland, W. S., and McGill, R. (1984b). The many faces of a scatterplot. *Journal of the American Statistical Association*, 79, 807–822.

Cook, D., and Swayne, D. F. (2007). *Interactive and Dynamic Graphics for Data Analysis: With R and GGobi*. New York: Springer.

Cook, R., and Wainer, H. (2012). A century and a half of moral statistics in the United Kingdom: Variations on Joseph Fletcher's thematic maps. *Significance*, 9(3), 31–36.

Cotes, R. (1722). *Aestimatio Errorum in Mixta Mathesis, per Variationes Planitum Trianguli Plani et Spherici*. (n.p.). Published in *Harmonia mensurarum*, Robert Smith, ed. Cambridge.

Dahmann, D. C. (2001). Presenting the nation's cultural geography [in census atlases]. Online at American Memory: Historical Collections for the National Digital Library Internet. http://memory.loc.gov/ammem/gmdhtml/census.html.

Delaney, J. (2012). *First X, Then Y, Now Z: An Introduction to Landmark Thematic Maps*. Darby, PA: Diane Publishing Co.

de Moivre, A. (1725). *Annuities upon Lives*. London: W. P. and Francis Fayram.

Descartes, R. (1637). La géométrie. In *Discours de la méthode*. Paris: Essellier. Appendix.

Diard, H. (1866). *Statistique morale de l'Angleterre et de la France par M. A-M. Guerry: Etudes sur cet ouvrage*. Paris: Baillière et fils.

Diard, H. (1867). *Discours de M. A. Maury et notices de MM. H. Diard et E. Vinet [On André-Michel Guerry]*. Baillière et fils. BNF: 8- LN27- 23721. I. Discours de M. Alfred Maury. II. Notice de M. H. Diard (*Journal d'Indre-et-Loire*, 13 Apr. 1866). III. Notice de M. Ernest Vinet (*Journal des débats*, 27 Apr. 1866).

Donoho, A. W., Donoho, D. L., and Gasko, M. (1988). Macspin: Dynamic graphics on a desktop computer. *IEEE Computer Graphics and Applications*, 8(4), 51–58.

Du Bois, W. E. B. (1900). The American Negro at Paris. *American Monthly Review of Reviews*, 22(5), 576.

Durkheim, E. (1897). *Le suicide*. Paris: Alcan. Trans. J. A. Spalding. Toronto: Collier-MacMillan, 1951.

Ekman, G. (1954). Dimensions of color vision. *Journal of Psychology*, 38, 467–474.

Emerson, J. W., Green, W. A., Schloerke, B., Crowley, J., Cook, D., Hofmann, H., and Wickham, H. (2013). The generalized pairs plot. *Journal of Computational and Graphical Statistics*, 22(1), 79–91.

Engel, E. (1857). Die productions- und consumtionsverhaltnisse des konigreichs sachsen. In *Die lebenkosten Belgischer arbeiter-familien*. Dresden: C. Heinrich, 1895.

Eyler, J. M. (1973). William Farr on the cholera: The sanitarian's disease theory and the statistician's method. *Journal of the History of Medicine and Allied Sciences*, 28(2), 79–100.

Eyler, J. M. (2001). The changing assessment of John Snow's and William Farr's cholera studies. *Soz Praventivmed*, 46(4), 225–232.

Farr, W. (1839). Letter to the registrar general. In *First Annual Report of the Registrar General*. London: HMSO.

Farr, W. (1852). *Report on the Mortality from Cholera in England, 1848–49*. London: HMSO.

Faure, F. (1918). The development and progress of statistics in France. In J. Koren, ed., *The History of Statistics: Their Development and Progress in Many Countries* (pp. 218–329). New York: Macmillan.

Fisherkeller, M. A., Friedman, J. H., and Tukey, J. W. (1974). PRIM-9: An interactive multidimensional data display and analysis system. Tech. Rep. SLAC-PUB-1408. Stanford Linear Accelerator Center, Stanford, CA.

Frère de Montizon, A. J. (1830). *Carte philosophique figurant la population de la France*. Paris, n.p.

Friedman, J. H., and Stuetzle, W. (2002). John W. Tukey's work on interactive graphics. *Annals of Statistics*, 30(6), 1629–1639.

Friedman, J. H., and Tukey, J. W. (1974). A projection pursuit algorithm for exploratory data analysis. *IEEE Transactions on Computers*, C-23(9), 881–890.

Friendly, M. (1991). *SAS System for Statistical Graphics*. 1st ed. Cary, NC: SAS Institute.

Friendly, M. (1999). Extending mosaic displays: Marginal, conditional, and partial views of categorical data. *Journal of Computational and Graphical Statistics*, 8(3), 373–395.

Friendly, M. (2002). Visions and re-visions of Charles Joseph Minard. *Journal of Educational and Behavioral Statistics*, 27(1), 31–51.

Friendly, M. (2005). Milestones in the history of data visualization: A case study in statistical historiography. In C. Weihs and W. Gaul, eds., *Classification: The Ubiquitous Challenge* (pp. 34–52). New York: Springer.

Friendly, M. (2007). A.-M. Guerry's *Moral Statistics of France:* Challenges for multivariable spatial analysis. *Statistical Science*, 22(3), 368–399.

Friendly, M. (2008a). A brief history of data visualization. In C. Chen, W. Härdle, and A. Unwin, eds., *Handbook of Computational Statistics: Data Visualization* (3:15–56). Heidelberg: Springer-Verlag.

Friendly, M. (2008b). The Golden Age of statistical graphics. *Statistical Science*, 23(4), 502–535.

Friendly, M. (2008c). La vie et l'oeuvre d'André-Michael Guerry (1802–1866). *Mémoires de l'Académie de Touraine*, 20. Read Feb. 8, 2008, Académie de Touraine.

Friendly, M., and Denis, D. (2000). The roots and branches of statistical graphics. *Journal de la Société Française de Statistique*, 141(4), 51–60. (Published in 2001).

Friendly, M., and Denis, D. (2005). The early origins and development of the scatterplot. *Journal of the History of the Behavioral Sciences*, 41(2), 103–130.

Friendly, M., and de Saint Agathe. (2012). André-Michel Guerry's *Ordonnateur Statistique*: The first statistical calculator? *American Statistician*, 66(3), 195–200.

Friendly, M., and Kwan, E. (2003). Effect ordering for data displays. *Computational Statistics and Data Analysis*, 43(4), 509–539.

Friendly, M., and Les Chevaliers des Albums de Statistique Graphique. (2020). Raiders of the Lost Tombs: The Search for Some Heroes of the History of Data Visualization, https://tinyurl.com/friendly-tombs.

Friendly, M., Monette, G., and Fox, J. (2013). Elliptical insights: Understanding statistical methods through elliptical geometry. *Statistical Science*, 28(1), 1–39.

Friendly, M., and Palsky, G. (2007). Visualizing nature and society. In J. R. Ackerman and R. W. Karrow, eds., *Maps: Finding Our Place in the World* (pp. 205–251). Chicago: University of Chicago Press.

Friendly, M., Sigal, M., and Harnanansingh, D. (2015). The Milestones Project: A database for the history of data visualization. In M. Kimball and C. Kostelnick, eds., *Visible Numbers: The History of Data Visualization*, chap. 10. London: Ashgate Press.

Friendly, M., Valero-Mora, P., and Ulargui, J. I. (2010). The first (known) statistical graph: Michael Florent van Langren and the "secret" of longitude. *American Statistician*, 64(2), 185–191.

Frost, R. (1979). *The Poetry of Robert Frost: The Collected Poems, Complete and Unabridged*. Ed. E. C. Latham. New York: Henry Holt & Co.

Funkhouser, H. G. (1936). A note on a tenth century graph. *Osiris*, 1, 260–262.

Funkhouser, H. G. (1937). Historical development of the graphical representation of statistical data. *Osiris*, 3(1), 269–405. Reprint. Bruges, Belgium: St. Catherine Press, 1937.

Galton, F. (1863a). A development of the theory of cyclones. *Proceedings of the Royal Society*, 12, 385–386.

Galton, F. (1863b). *Meteorographica, or Methods of Mapping the Weather*. London: Macmillan. BL: Maps.53.b.32.

Galton, F. (1866). On the conversion of wind-charts into passage-charts. *Philosophical Magazine*, 32, 345–349.

Galton, F. (1886). Regression towards mediocrity in hereditary stature. *Journal of the Anthropological Institute of Great Britain and Ireland*, 15, 246–263.

Galton, F. (1889). *Natural Inheritance*. London: Macmillan.

Galton, F. (1890). Kinship and correlation. *North American Review*, 150, 419–431.

Galton, S. T. (1813). *A Chart, Exhibiting the Relation between the Amount of Bank of England Notes in Circulation, the Rate of Foreign Exchanges, and the Prices of Gold and Silver Bullion and of Wheat*. London: Johnson & Co.

Gelman, A., and Unwin, A. (2013). Infovis and statistical graphics: Different goals, different looks. *Journal of Computational and Graphical Statistics*, 22(1), 2–28.

Geneen, H. (1985). *Managing*. New York: Avon Books.

General Register Office. (1852). *Report on the Mortality of Cholera in England, 1848–49*. London: W. Clowes and Sons, for Her Majesty's Stationery Office. Written by William Farr.

Gilbert, E. W. (1958). Pioneer maps of health and disease in England. *Geographical Journal*, 124, 172–183.

Goethe, J. W. (2018). *The Essential Goethe*. Ed. M. Bell. Princeton, NJ: Princeton University Press.

Guerry, A.-M. (1829). Tableau des variations météorologique comparées aux phénomènes physiologiques, d'aprés les observations faites à l'obervatoire royal, et les recherches statistique les plus récentes. *Annales d'hygiène publique et de médecine légale*, 1, 228–237.

Guerry, A.-M. (1833). *Essai sur la statistique morale de la France*. Paris: Crochard.

Hacking, I. (1990). *The Taming of Chance*. Cambridge: Cambridge University Press.

Halley, E. (1686). On the height of the mercury in the barometer at different elevations above the surface of the earth, and on the rising and falling of the mercury on the change of weather. *Philosophical Transactions*, 16, 104–115.

Halley, E. (1693). An estimate of the degrees of mortality of mankind, drawn from curious tables of the births and funerals at the city of Breslaw, with an attempt to ascertain the price of annuities on lives. *Philosophical Transactions*, 17, 596–610.

Halley, E. (1701). *The description and uses of a new, and correct sea-chart of the whole world, shewing variations of the compass*. London: published by author.

Hamming, R. W. (1962). *Numerical Methods for Scientists and Engineers*. New York: McGraw-Hill.

Hankins, T., and Silverman, R. (1999). *Instruments and the Imagination*. Princeton, NJ: Princeton University Press.

Hankins, T. L. (1999). Blood, dirt, and nomograms: A particular history of graphs. *Isis*, 90, 50–80.

Hankins, T. L. (2006). A "large and graceful sinuosity": John Herschel's graphical method. *Isis*, 97, 605–633.

Harari, Y. N. (2015). *Sapiens: A Brief History of Humankind*. New York: Harper.

Hartigan, J. A. (1975). Printer graphics for clustering. *Journal of Statistical Computing and Simulation*, 4, 187–213.

Hayes, J. N. (2005). *Epidemics and Pandemics: Their Impacts on Human History*. Santa Barbara, CA: ABC-CLIO.

Herschel, J. F. W. (1833a). III. Micrometrical measures of 364 double stars with a 7-feet equatorial acromatic telescope, taken at Slough, in the years 1828, 1829, and 1830. *Memoirs of the Royal Astronomical Society*, 5, 13–91. Communicated Feb. 8, 1831. Read May 13 and June 10, 1831.

Herschel, J. F. W. (1833b). On the investigation of the orbits of revolving double stars: Being a supplement to a paper entitled "micrometrical measures of 364 double stars." *Memoirs of the Royal Astronomical Society*, 5, 171–222.

Herschel, J. F. W. (1860). *Outlines of Astronomy*. Philadelphia, PA: Blanchard & Lea.

Hilts, V. L. (1975). *A Guide to Francis Galton's English Men of Science*, Vol. 65, Part 5. Philadelphia, PA: Transactions of the American Philosophical Society.

Hoff, H. E., and Geddes, L. A. (1962). The beginnings of graphic recording. *Isis*, 53, pt. 3, 287–324.

Hofmann, H. (2007). Interview with a centennial chart. *Chance*, 20(2), 26–35.

Howard, L. and Geological Society of London. (1847). *Barometrographia: Twenty Years' Variation of the Barometer in the Climate of Britain, Exhibited in Autographic Curves, with the Attendant Winds and Weather, and Copious Notes Illustrative of the Subject*. Richard and John E. Taylor. https://books.google.ca/books?id=NfYKQgAACAAJ.

International Statistical Congress. (1858). Emploi de la cartographic et de la méthode graphique en général pour les besoins spéciaux de la statistique. In *Proceedings* (pp. 192–197). Vienna. 3rd Session, August 31–September 5, 1857.

Jevons, W. S. (1863). *A Serious Fall in the Value of Gold Ascertained, and Its Social Effects Set Forth*. London: Edward Stanford.

Johnson, S. (2006). *The Ghost Map: The Story of London's Most Terrifying Epidemic—And How It Changed Science, Cities, and the Modern World*. New York: Riverhead Books.

Kane, T., and Scher, M. (1969). A dynamical explanation of the falling cat phenomenon. *International Journal of Solids and Structures*, 5(7), 663–670.

Koch, T. (2000). *Cartographies of Disease: Maps, Mapping, and Medicine*. Redlands, CA: ESRI Press.

Koch, T. (2004). The map as intent: Variations on the theme of John Snow. *Cartographica*, 39(4), 1–14.

Koch, T. (2011). *Disease Maps: Epidemics on the Ground*. Chicago: University of Chicago Press.

Koch, T. (2013). Commentary: Nobody loves a critic: Edmund A Parkes and John Snow's cholera. *International Journal of Epidemiology*, 42(6), 1553.

Kopf, E. W. (1916). Florence Nightingale as statistician. *Publications of the American Statistical Association*, 15(116), 388–404.

Kruskal, W. (1978). Taking data seriously. In Y. Elkana, J. Lederberg, R. Morton, A. Thackery, and H. Zuckerman, eds., *Toward a Metric of Science: The Advent of Science Indicators* (pp. 139–169). New York: Wiley.

Kuhn, T. S. (1970). *The Structure of Scientific Revolutions*. Chicago: University of Chicago Press.

Lalanne, L. (1844). *Abaque, ou compteur univsersel, donnant á vue á moins de 1/200 près les résultats de tous les calculs d'arithmétique, de geometrie et de mécanique practique*. Paris: Carilan-Goery et Dalmont.

Lalanne, L. (1845). Appendice sur la representation graphique des tableaux météorologiques et des lois naturelles en général. In L. F. Kaemtz, ed., *Cours complet de météorologie* (pp. 1–35). Paulin. Trans. and annotated C. Martins.

Lalanne, L. (1879). *Methodes graphiques pour l'expression des lois empiriques ou mathematiques à trois variables.* Paris: Imprimerie Nationale.

Lallemand, C. (1885). *Les abaques hexagonaux: Nouvelle méthode générale de calcul graphique, avec de nombreux exemples d'application.* Paris: Ministère des travaux publics, Comité du nivellement général de la France.

Lambert, J. H. (1765). Theorie der zuverlässigkeit. In *Beyträge zum gebrauche der mathematik and deren anwendungen* (1:424–488). Berlin: Verlage des Buchladens der Realschule.

Lambert, J. H. (1779). *Pyrometrie; oder, vom maasse des feuers und der wärme mit acht kupfertafeln.* Berlin: Haude & Spener.

Lemann, N. (1991). *The Promised Land: The Great Black Migration and How It Changed America.* New York: Knopf.

Levasseur, É. (1885). La statistique graphique. *Journal of the Statistical Society of London*, 50, 218–250.

Lexis, W. (1875). Einleitung in der theorie der bevölkerungsstatistik.

Maclean, N. (1992). *Young Men and Fire.* Chicago: University of Chicago Press.

Mandea, M., and Mayaud, P.-N. (2004). Guillaume Le Nautonier, un precurseur dans l'histoire du géomagnétisme magnetism. *Revue d'histoire des sciences*, 57(1), 161–174.

Marey, É. J. (1873). *La machine animale, locomotion terestre et aérienne.* Paris: Baillière.

Marey, E. J. (1878). *La méthode graphique dans les sciences expérimentales et principalement en physiologie et en médecine.* Paris: G. Masson.

Marey, E. J. (1885). *La méthode graphique.* Paris: Boulevard Saint Germain et rue de l'Eperon.

Marey, E. J. (1894). Des mouvements que certains animaux exécutent pour retomber sur leurs pieds, lorsqu'ils sont précipités d'un lieu élevé. *Comptes rendus de l'Academie des Sciences*, 119, 714–717.

Marsaglia, G. (1968). Random numbers fall mainly in the planes. *Proceedings of the National Academy of Sciences*, 61(1), 25–28.

Marshall, A. (1885). On the graphic method of statistics. *Journal of the Royal Statistical Society*, 50 (Jubilee volume), 251–260. Read at the International Statistical Congress, held at the Jubilee of the Statistical Society of London, June 23, 1885.

Maurage, P., Heeren, A., and Pesenti, M. (2013). Does chocolate consumption really boost Nobel Award chances? The peril of over-interpreting correlations in health studies. *Journal of Nutrition*, 143(6), 931–933.

McDonald, J. A. (1982). Interactive graphics for data analysis. Ph.D. thesis, Stanford University.

Messerli, F. H. (2012). Chocolate consumption, cognitive function, and Nobel laureates. *New England Journal of Medicine*, 367(16), 1562.

Minard, C. J. (1856). *De la chute des ponts dans les grandes crues.* Paris: E. Thunot et Cie. ENPC: 4-4921/C282.

Minard, C. J. (1862a). *Carte figurative et approximative représentant pour l'année 1858 les émigrants du globe*. Regnier et Dourdet. ENPC: Fol 10975.

Minard, C. J. (1862b). *Des tableaux graphiques et des cartes figuratives*. Paris: E. Thunot et Cie. ENPC: 3386/C161, BNF: Tolbiac, V-16168.

Monmonier, M. (1991). *How to Lie with Maps*. Chicago: The University of Chicago Press.

Monmonier, M. (1999). *Air Apparent: How Meteorologists Learned to Map, Predict and Dramatize Weather*. Chicago: University of Chicago Press.

Mumford, L. (2000). *Art and Technics (The Bampton Lectures in America)*. New York: Columbia University Press.

Murray, L. L., and Bellhouse, D. R. (2017). How was Edmond Halley's map of magnetic declination (1701) constructed? *Imago Mundi*, 69(1), 72–84.

Musée National d'Art Moderne. (1991). *André Breton. La beautéconvulsive*. Paris: Editions du Centre Pompidou. Exhibition Catalog.

Muybridge, E. (1878). The Horse in motion. "Sallie Gardner," owned by Leland Stanford; running at a 1:40 gait over the Palo Alto track, 19th June 1878. Cabinet cards, Morses' Gallery, San Francisco.

Newton, C. M. (1978). Graphics: From alpha to omega in data analysis. In P. C. C. Wang, ed., *Graphical Representation of Multivariate Data*. New York: Academic Press. Proceedings of the Symposium on Graphical Representation of Multivariate Data, Naval Postgraduate School, Monterey CA, Feb. 24, 1978.

Nightingale, F. (1858). *Notes on Matters Affecting the Health, Efficiency, and Hospital Administration of the British Army*. London: Harrison and Sons. Presented by request to the Secretary of State for War.

Nightingale, F. (1859). *A Contribution to the Sanitary History of the British Army during the Late War with Russia*. London: John W. Parker and Son. UCL: UH258 1853.C73.

O'Connor, J. J., and Robertson, E. F. (1997). Longitude and the académie royale. MacTutor History of Mathematics. http://www-groups.dcs.st-and.ac.uk/~history/PrintHT/Longitude1.html.

Oresme, N. (1482). *Tractatus de latitudinibus formarum*. Padova.

Palsky, G. (1996). *Des chiffres et des cartes: Naissance et développement de la cartographie quantitative française au XIXe siècle*. Paris: Comité des Travaux Historiques et Scientifiques (CTHS).

Palsky, G. (1999). The debate on the standardization of statistical maps and diagrams (1857–1901). *Cybergeo*, n.v.(65). Retrieved from http://cybergeo.revues.org/148.

Parent-Duchâtelet, A. J. B. (1836). *De la prostitution dans la ville de Paris*. Bruxelles: Dumont.

Parkes, E. A. (1855). Review: *Mode of communication of cholera* by John Snow. *British and Foreign Medico-Churgical Review*, 15, 449–456. Reprinted in *Int. J. Epidemiol*, 42(6), 1543–1552.

Pearson, K. (1901). On lines and planes of closest fit to systems of points in space. *Philosophical Magazine*, 6(2), 559–572.

Pearson, K. (1914–1930). *The Life, Letters and Labours of Francis Galton*. 4 vols. Cambridge: Cambridge University Press.

Pearson, K. (1920). Notes on the history of correlation. *Biometrika*, 13, 25–45.

Perozzo, L. (1881). Stereogrammi demografici—Seconda memoria dell'ingegnere luigi perozzo. *Annali di Statistica*, 22, 1–20.

Petty, W. (1690). *Political Arithmetick*. 3rd ed. London: Robert Clavel.

Phillips, A. W. H. (1958). The relation between unemployment and the rate of change of money wage rates in the United Kingdom, 1861–1957. *Economica*, n.s. 25(2), 283–299.

Playfair, W. (1786). *Commercial and Political Atlas: Representing, by Copper-Plate Charts, the Progress of the Commerce, Revenues, Expenditure, and Debts of England, during the Whole of the Eighteenth Century*. London: Debrett, Robinson, and Sewell. Reprinted 2005 in H. Wainer and I. Spence, eds., *The Commercial and Political Atlas and Statistical Breviary*. Cambridge: Cambridge University Press.

Playfair, W. (1801). *Statistical Breviary; Shewing, on a Principle Entirely New, the Resources of Every State and Kingdom in Europe*. London: Wallis. Reprinted 2005 in H. Wainer and I. Spence, eds., *The Commercial and Political Atlas and Statistical Breviary*. Cambridge: Cambridge University Press.

Playfair, W. (1805). *An Inquiry into the Permanent Causes of the Decline and Fall of Powerful and Wealthy Nations . . . Designed to Shew How the Prosperity of the British Empire May Be Prolonged*. London: Greenland and Norris.

Playfair, W. (1821). *Letter on Our Agricultural Distresses, Their Causes and Remedies; Accompanied with Tables and Copperplate Charts Shewing and Comparing the Prices of Wheat, Bread and Labour, from 1565 to 1821*. London: W. Sams. BL: 8275.c.64.

Plot, R. (1685). A letter from Dr. Robert Plot of Oxford to Dr. Martin Lister of the Royal Society concerning the use which may be made of the following history of the weather made by him at Oxford throughout the year 1864. *Philosophical Transactions*, 169, 930–931.

Poole, K. T. (2005). *Spatial Models of Parliamentary Voting (Analytical Methods for Social Research)*. Cambridge: Cambridge University Press.

Popper, W. (1951). *The Cairo Nilometer: Studies in Ibn Taghrî Birdî's Chronicles of Egypt: I*. Publications in Semitic Philology, vol. 12. Berkeley: University of California Press.

Priestley, J. (1765). *A Chart of Biography*. London: n.p. BL: 611.I.19.

Priestley, J. (1769). *A New Chart of History*. London: Thomas Jeffreys. BL: Cup.1250.e.18 (1753).

Radzinowicz, L. (1965). Ideology and crime: The deterministic position. *Columbia Law Review*, 65(6), 1047–1060.

Reaven, G., and Miller, R. (1968). Study of the relationship between glucose and insulin responses to an oral glucose load in man. *Diabetes*, 17(9), 560–569. American Diabetes Association. http://diabetes.diabetesjournals.org/content/17/9/560.full.pdf.

Reaven, G. M., and Miller, R. G. (1979). An attempt to define the nature of chemical diabetes using a multidimensional analysis. *Diabetologia*, 16, 17–24.

Rendgen, S. (2018). *The Minard System: The Complete Statistical Graphics of Charles-Joseph Minard*. New York: Princeton Architectural Press.

Robinson, A. H. (1982). *Early Thematic Mapping in the History of Cartography*. Chicago: University of Chicago Press.

Rosenberg, D., and Grafton, A. (2010). *Cartographies of Time: A History of the Timeline*. New York: Princeton Architectural Press.

Rosling, H., and Johansson, C. (2009). Gapminder: Liberating the x-axis from the burden of time. *Statistical Computing and Statistical Graphics Newsletter*, 20(1), 4–7.

Rubin, E. (1943). The place of statistical methods in modern historiography. *American Journal of Economics and Sociology*, 2(2), 193–210.

Schwabe, H. (1872). Theorie der graphischen darstellungen. In P. Séménov, ed., *Proceedings of the International Statistical Congress*, 8th Session, Pt. 1 (pp. 61–73). St. Petersburg: Trenké & Fusnot.

Shiode, N., Shiode, S., Rod-Thatcher, E., Rana, S., and Vinten-Johansen, P. (2015). The mortality rates and the space-time patterns of John Snow's cholera epidemic map. *International Journal of Health Geographics*, 14(1), 21.

Slocum, T. A., McMaster, R. B., Kessler, F. C., and Howard, H. H. (2008). *Thematic Cartography and Geographic Visualization*. New York: Pearson / Prentice Hall.

Snow, J. (1849a). *On the Mode of Communication of Cholera*. 1st ed. London: J. Churchill.

Snow, J. (1849b). On the pathology and mode of transmission of cholera. *Medical Gazette and Times*, 44, 745–752, 923–929.

Snow, J. (1855). *On the Mode of Communication of Cholera*. 2nd ed. London: J. Churchill.

Sobel, D. (1996). *Longitude: The True Story of a Lone Genius Who Solved the Greatest Scientific Problem of His Time*. New York: Penguin.

Somerhausen, H. (1829). Carte figurative de l'instruction populaire de pay bas. Bruxelles. n.p.

Spence, I. (2006). William Playfair and the psychology of graphs. In *Proceedings of the American Statistical Association, Section on Statistical Graphics* (pp. 2426–2436). Alexandria, VA: American Statistical Association.

Spence, I., Fenn, C. R., and Klein, S. (2017). Who is buried in Playfair's grave? *Significance*, 14(5), 20–23.

Spence, I., and Garrison, R. F. (1993). A remarkable scatterplot. *American Statistician*, 47(1), 12–19.

Spence, I., and Wainer, H. (1997). William Playfair: A daring worthless fellow. *Chance*, 10(1), 31–34.

Spence, I., and Wainer, H. (2005). William Playfair and his graphical inventions: An excerpt from the introduction to the republication of his *Atlas* and *Statistical Breviary*. *American Statistician*, 59(3), 224–229.

Stamp, J. (1929). *Some Economic Factors in Modern Life*. London: P. S. King & Son.

Stanton, J. M. (2001). Galton, Pearson, and the peas: A brief history of linear regression for statistics instructors. *Journal of Statistics Education*, 9(3).

Statistischen Bureau. (1897). *Graphisch-statistischer Atlas der Schweiz (Atlas Graphique et Statistique de la Suisse)*. Departments des Innern, Bern: Buchdruckeri Stämpfli & Cie.

Statistischen Bureau. (1914). *Graphisch-statistischer Atlas der Schweiz (Atlas Graphique et Statistique de la Suisse)*. Bern: LIPS & Cie.

Stigler, S. M. (1980). Stigler's law of eponomy. *Transactions of the New York Academy of Sciences*, 39, 147–157.

Stigler, S. M. (1986). *The History of Statistics: The Measurement of Uncertainty before 1900*. Cambridge, MA: Harvard University Press.

Stigler, S. M. (1989). Francis Galton's account of the invention of correlation. *Statistical Science*, 4(2), 73–79.

Stigler, S. M. (1999). *Statistics on the Table: The History of Statistical Concepts and Methods*. Cambridge, MA: Harvard University Press.

Stigler, S. M. (2016). *The Seven Pillars of Statistical Wisdom*. Cambridge, MA: Harvard University Press.

Süssmilch, J. P. (1741). *Die göttliche Ordnung in den Veränderungen des menschlichen Geschlechts, aus der Geburt, Tod, und Fortpflantzung*. Germany: n.p. Published in French translation as *L'ordre divin. dans les changements de l'espèce humaine, démontré par la naissance, la mort et la propagation de celle-ci*. Trans. Jean-Marc Rohrbasser. Paris: INED.

Sutherland, I. E. (1963). Sketchpad: A man-machine graphical communication system. Ph.D. thesis, MIT. Available as Computer Laboratory Technical Report, University of Cambridge UCAM-CL-TR-574, September 2003.

Thrower, N. J. W., ed. (1981). *The Three Voyages of Edmond Halley in the* Paramore *1698–1701*. London: Hakluyt Society. 2nd series, vol. 156–157 (2 vols.).

Tufte, E. R. (1983). *The Visual Display of Quantitative Information*. Cheshire, CT: Graphics Press.

Tufte, E. R. (1990). *Envisioning Information*. Cheshire, CT: Graphics Press.

Tufte, E. R. (1997). *Visual Explanations: Images and Quantities, Evidence and Narrative*. Cheshire, CT: Graphics Press.

Tufte, E. R. (2006). *Beautiful Evidence*. Cheshire, CT: Graphics Press.

Tukey, J. W. (1962). The future of data analysis. *Annals of Mathematical Statistics*, 33(1), 1–67.

Tukey, J. W. (1972). Some graphic and semigraphic displays. In T. A. Bancroft, ed., *Statistical Papers in Honor of George W. Snedecor* (pp. 292–316). Ames: Iowa State University Press.

Tukey, J. W. (1977). *Exploratory Data Analysis*. Reading, MA: Addison Wesley.

Tukey, J. W., and Tukey, P. A. (1985). Computer graphics and exploratory data analysis: An introduction. In *Proceedings of the Sixth Annual Conference and Exposition: Computer Graphics85*. Fairfax, VA: National Computer Graphics Association.

van Langren, M. F. (1644). *La Verdadera Longitud por Mar y Tierra*. Antwerp: n.p. BL: 716.i.6.(2.); BeNL: VB 5.275 C LP.

Vauthier, L.-L. (1874). Note sur une carte statistique figurant la répartition de la population de Paris. *Comptes rendus des séances de L'Académie des Sciences*, 78, 264–267. ENPC: 11176 C612.

Velleman, P. F., and Velleman, A. Y. (1985). *Data Desk Handbook*. Ithaca, NY: Data Description.

Venn, J. (1880). On the diagrammatic and mechanical representation of propositions and reasonings. *London, Edinburgh, and Dublin Philosophical Magazine and Journal of Science*, 9, 1–18.

Wainer, H. (1996). Why Playfair? *Chance*, 9(2), 43–52.

Wainer, H. (2005). *Graphic Discovery: A Trout in the Milk and Other Visual Adventures*. Princeton, NJ: Princeton University Press.

Wainer, H., and Spence, I. (1997). Who was Playfair? *Chance*, 10(1), 35–37.

Walker, F. A. (1874). *Statistical Atlas of the United States, Based on the Results of Ninth Census, 1870, with Contributions from Many Eminent Men of Science and Several Departments of the [Federal] Government*. New York: Julius Bien.

Wallis, H. M., and Robinson, A. H. (1987). *Cartographical Innovations: An International Handbook of Mapping Terms to 1900*. Tring, UK: Map Collector Publications.

Watt, J. (1822). Notice of his important discoveries in powers and properties of steam. *Quarterly Journal of Science, Literature and the Arts*, 11, 343–345.

Wauters, A. (1891). LANGREN (Michel-Florent VAN). *Biographie Nationale*, E. Brulant. Academie Royal de Belgique, 11.

Wauters, A. (1892). Michel-Florent van Langren. *Ciel et Terre*, 12, 297–304.

Whitaker, E. A. (2003). *Mapping and Naming the Moon: A History of Lunar Cartography and Nomenclature*. Cambridge: Cambridge University Press.

Whitehead, M. (2000). William Farr's legacy to the study of inequalities in health. *Bulletin of the World Health Organization*, 78(1), 86–87. https://www.who.int/bulletin/archives/78(1)86.pdf.

Zeuner, G. (1869). *Abhandlungen aus der mathematischen statistik*. Leipzig: Verlag von Arthur Felix. BL: 8529.f.12.

Acknowledgments

Any project that has gestated as long as this one must, of necessity, also accrue enormous debts to mentors, collaborators, inspirators, students, and friends who have contributed to our thinking and our work. Principal among them are David Hoaglin and Stephen Stigler.

Dave took on the role of an amicus curiae for the first draft of this book, going through our text in thoughtful detail and improving both what was presented and the clarity of the presentation. We have not words enough to thank him.

Steve Stigler, the world's preeminent historian of statistics, was an invaluable resource at the other end of our emails. He provided facts, direction, and corrections, and embarrassingly often he would dig into his extraordinary personal collection of historical documents to share a high-quality scan of one figure or another, often appending the best citation should we need it. It is not hyperbole to state that without the help of these two scholars this book would be very much diminished.

Either formally or informally, we have been fortunate to have been mentored by people of both wisdom and kindness: Jacques Bertin, Albert Biderman, Harold Gullisken, George Miller, Frederick Mosteller, Peter Ornstein, Edward Tufte, and John Wilder Tukey.

This book also owes a great debt to a group of scholars (and friends) known collectively as Les Chevaliers des Albums de Statistique Graphique. It was formed initially to purchase collectively and celebrate a collection of historical volumes produced by the Ministry of Public Works in France, 1879–1899, perhaps the most exquisite sampler of graphical methods ever produced. The founding members of this group included Antoine de Falguerolles, Gilles Palsky, Ian Spence, Antony Unwin, Forrest Young, and Michael Greenacre. Since then, several historians of science (Stephen Stigler, Ted Porter, Tom Koch) and others with an interest in the history of data visualization have been invited to join; most recently: Raymond J. Andrews, David Rumsey, and Sandra Rendgen. In writing this book, Les Chevaliers have been extraordinarily generous with their expertise and insight.

Invaluable help, support, and advice were provided by our colleagues, friends, and students. Prominent among them are William Berg, Henry Braun, Rob Cook, Cathy

Durso, Richard Feinberg, Peter Li, Ernest Kwan, Sam Palmer, Jim Ramsay, Matthew Sigal, Linda Steinberg, David Thissen, Paul Velleman, and Leland Wilkinson.

Work on the Milestones Project and the history of data visualization over many years was supported by Touraj Amiri, Daniel J. Denis, Matthew Dubins, Derek Harmanansingh, Joo Ann Lee, Pere Millán, Carolina Patryluk, Gustavo Vierira, and Justeena Zaki-Azat.

We are delighted to be able to acknowledge the help and encouragement of a very special kind that were provided by editors past and present. Prominent among these are Cambridge's Lauren Cowles and Princeton's Vickie Kearns, who put their publishers' money where our mouths were in earlier books.

Most relevant for this book are Harvard's Thomas Embree LeBien and Janice Audet, whose guidance throughout the writing and publication process shaped our rough manuscript into the finished volume you now hold in your hand. Additionally, we are grateful to Emeralde Jensen-Roberts, Stephanie Vyce, and the editorial and production staff at Harvard University Press and also to Melody Negron at Westchester Publishing Services for lending their special skills to allow us to more nearly say what we meant and show it in a visually appealing way. Outstanding among those are Emeralde Jensen-Roberts and Stephanie Vyce for working helpfully with us on production and intellectual property issues.

Index

Page numbers in italics refer to illustrations.